T0319016

The organisation of transactions

The organisation of transactions

Studying supply networks using gaming simulation

Sebastiaan Meijer

International chains and network series – Volume 6

Wageningen Academic
P u b l i s h e r s

ISBN 978-90-8686-102-6
ISSN 1874-7663

First published, 2009

Wageningen Academic Publishers
The Netherlands, 2009

Preface and acknowledgements

The romantic ideal of doing a PhD is to dive into a topic of research and write down interesting conclusions in a thesis without getting distracted. This book is the result of a longer and more differentiated process, a process that was needed to help me to write coherently.

While my PhD project started six years ago, the root of this book is to be found in the Chaingame project: the topic of my Master thesis. The Chaingame was a computerised gaming simulation for studying transactions in a supply network. Since then a lot has happened, but transactions in a supply network are still the topic of research. The Chaingame was an idea of my later supervisor and co-promotor Gert Jan Hofstede.

In my PhD project I am indebted to my *promotores* Onno Omta and George Beers. The fact that this book is now finished illustrates your professionalism. Onno, you have been of great help with the data analysis. You were a benchmark to satisfy when scientific results had to be presented. I learned from your drive to score interesting publications. George, you have been the methodological conscience of this project. Your clear thinking made it possible to seek the unknown research method in a broad theoretical field. Thank you for staying on board when your work situation changed.

The two studies presented in this book are both the work of a group of people. For the Trust and Tracing Game I benefited from the work of four students who worked on the first data analysis during the design cycle (Kim Dooper, Femke van der Geer, Linda Haan and Elise Keurentjes). The enormous number of envelopes could be digitalised thanks to my friend in research Tim Verwaart who arranged some weeks of work capacity. The multi-agent simulation started together with Tim too. Several students worked on this. Martijn Haarman, Danny Kortekaas and Hans Klaverweijden contributed to early versions. The real progress was made when Tim came up with a brilliant, and silent, M.Sc. student called Dmytro Tykhonov, and with him Catholijn Jonker. I am honoured that I could work with such bright people and hopefully will continue to do so, together with Koen Hindriks and Gert Jan Hofstede. Hummelo forever!

Several people made it possible to conduct sessions with larger numbers of students. At Wageningen University Jacques Trienekens gave me access to students Supply Chain Management. At INHOLLAND Delft, Kees Vermeulen was the one who dared to give me time and space to conduct my experiments. Adrie Beulens arranged contact with the groups from Purdue University. Olaf van Kooten arranged contacts with the tomato networks in Limburg.

The Mango Chain Game was a joint cooperation between Guillermo Zúñiga-Arias and me. Soon we found Sietse Sterrenburg as a key contributor to the game design. Guillermo and Sietse, I still have fond memories of my weeks in Costa Rica together with you. The focused teamwork, combined with enjoyable trips and discussions in a hot spring made for an unforgettable experience. Thank you! The hidden architect of this project was Ruerd Ruben. Thank you Ruerd for believing we would achieve something with the game sessions.

For the past eight years I have been a student, colleague and guest of the former Information Technology Group at Wageningen University. I want to thank the people in this group for showing interest during these years. I enjoyed the tolerant atmosphere, especially during lunches in the previous building. May the new combination in LDI bring new energy. I want to mention three people specifically. Yuan Li, you were the most laughing roommate I ever had. Sorry for the mess when I was making game materials. Jan Ockeloen, you were of great help when I was still working on computer games, and later when making game materials. I hope you will enjoy retirement and think only about the good years at the Computechnion. Mark Kramer, I said it before and repeat it here: you have been an enormous influence on the way that I see the world. You taught me how to think in abstractions and generalisations, in a way showing a deep fascination for programming. You are a role model as a teacher.

Most of all, I want to mention my supervisor and co-promotor Gert Jan Hofstede. Dear Gert Jan, you believed in the potential of that young solistic tall student with his serious ups and downs. I believed in your ideas about research with gaming simulations. You never ceased to encourage me to improve. I will never get up to par with your language skills but have learned so much about writing from your constructive criticism. You have been a friend when possible, a mentor when I needed it, and a critic when I deserved it. You have been responsible for giving me inspiration to finish my M.Sc. and to enrol in a PhD project. You pulled me back when I ran away to music and pushed me to climb the hill. It is impossible to thank you enough, so I will stick to a simple 'thank you'.

As of August 2008, I started in a new position as assistant professor at the faculty of technology, policy and management of Delft University of Technology. I never thought I would continue in science after my M.Sc. and again never felt attracted to science during my PhD. Live music was my biggest passion, and as a sound engineer I could contribute to this passion. The last year of this project however I saw the light and discovered the joys of writing. I have been doing sound for live bands and theatre for the last ten years, with a semi-permanent touring schedule during the last three years. The joy found in science made me decide to make music the large hobby that it was six years ago, and to concentrate on my new position. Parts of this book have been written on the road and figures were drawn in dressing rooms of theatres. To all my friends in music I want to say thank you for the good times. The speed of writing this book has suffered from my distraction with music, but the book wouldn't even have existed without sound as creative way to free new energy.

Last of all there are the people in my private life who observed my progress in research from the side lines. To my father and mother I want to say that I finally finished my 'werkstuk'. *Thuiskomen was altijd fijn om te praten of gewoon even thuis te zijn.* Sander, Els and Auke Jan: we are so different and yet so similar. Let's stay connected!

I hope that the reader will enjoy this book. The numerous sessions with gaming simulations that form the empirical basis of this book were a joy to conduct. I hope that the enthusiasm of the participants who inspired me to continue can be felt while reading.

Utrecht, 2008-11-02

Sebastiaan Meijer

Table of contents

List of tables

The organisation of transactions

List of figures

1. Introduction

Walking into a local grocery store or the fruit and vegetable section of a supermarket is like taking a look at different continents. In the modern world the food we eat, the equipment we use and clothes we wear are sourced from all over the world. Outsourcing of labour intensive production to emerging economies and consumer preference for fresh products throughout the year are just two of the drivers of the globalisation of production. Between the production and consumption each product has been traded numerous times. In the case of multinationals like Philips, Intel and Dole, one large firm that manufactures or grows, packs and distributes can control the whole process. In the case of most food products, however, large numbers of products are produced, traded and resold by independent firms. Food supply networks span a whole series of firms from grower to consumer. Depending on the product traded and the market in which it is consumed, the grower and consumer can be located in countries thousands of kilometres apart, passing several companies in the supply network. Still the food needs to get to the consumer in perfect condition. Transactions need to be made between the subsequent companies that trade the product. The way in which the transactions are organised can be any mix of the following three mechanisms: *market*, *network* and *hierarchy* (Powell, 1990). This book concentrates on market and network. The market mechanism uses the price as a control mechanism, whereas the network mechanism uses trust and reputation.

1.1 Study domain

This book starts from the perspective of the trade agent: the individual trader for a company. Traders have personal characteristics and have relationships with each other. The relationship that two traders have could make them more willing to deal with each other than with another trader (Hofstede, 2004). Traders can be from different countries and cultures and may not understand each other's signs of good will. Some traders have a large social network, while others may know just a few other traders. Personal friendships may limit the choice of business partners as well. These characteristics and relationships will affect the trade between companies and thus the organisation of transactions in the supply network. Lazzarini *et al.* (2001) identified the value of these variables and placed them in the bucket category *social structure* as one of the sources of value for supply networks. The social structure category contains variables related to interpersonal and business relationships and the norms and values in the supply network as a whole. Trust between traders is the most prominent interpersonal variable. Business relations are expressed in a level of embeddedness as a measure of the density and the strength of ties between businesses (Granovetter, 1985). Norms and values in a network can be related to the culture traders come from (Hofstede, 2004). The concept of social structure is covered by a broad range of theories. Because of its width (discussed later in Section 4.2) this book concentrates on the major variables of personal relations (trust), business relations (embeddedness) and cultural influences (norms and values).

This book studies the organisation of transactions in supply networks. More specifically it investigates the influence of social structure on network and/or market mode of organisation. The research question where this started is therefore the following:

> *What is the influence of social structure on the organisation of transactions in supply networks?*

This book is part of a research stream that focuses on the performance of supply chains and networks. The concept of the supply chain emerged in the 1980s to stress that the optimisation of production and distribution expands across the borders of the firm (Simchi-Levi *et al.*, 2000). The focus moved from in-company improvements to the full supply chain from raw materials to the end consumer. Logistics was the first topic to be analysed. During the 1990s the transactions and communication between firms became major topics of research, also incorporating layers of different suppliers and buyers and competing companies (Omta *et al.*, 2001; Zylberstajn, 2004). This view focuses not only on the supply chain with its product flow, but on the whole network of companies that operate in (parallel or connected) chains to create a product, the so-called supply network ((Lazzarini *et al.*, 2001). This book uses the term 'supply chains and networks' as one term, and uses 'supply network' as the short version of it. The term supply chain is used only when the sequential flow of goods is specifically important. Section 4.1 discusses the supply chain and network theory.

While trading, traders incur costs. Transaction costs are the costs associated with doing business that do not add value to the product. New institutional economics (Williamson, 1996) assumes that the mode of organisation (the mix of market, network or hierarchy) used is that with the lowest transaction costs associated for the particular setting of the supply network. Minimising transaction costs leads to a better performing network. A better performing network delivers more value to the consumer and more profits for the companies in the network.

Trade in supply networks has four sources of transaction costs: searching, bargaining, monitoring and enforcing (Williamson 1979; Holloway *et al.*, 2000). *Searching* for possible trade partners should ideally be minimised as the network partners know each other well. However, to avoid lock-in it should not be too difficult to contact a new trade partner. Norms and values could prohibit this. Repeated *bargaining* between partners who continue business with each other should ideally be avoided, for instance using long-term agreements. *Monitoring* of agreements should ideally be unnecessary when you can trust your network partners, as should be the case for *enforcement* of broken agreements. However, studies of supply network projects (Van der Vorst, 2004; a.o.) show that many networks function sub-optimally. Examples are companies that feel that others are taking advantage of them and companies that do not trust each other. Section 4.2 discusses the relationship between the four sources of transaction costs and the modes of organisation. It shows that depending on the mode of organisation the four sources appear in different ways. Analytically, the four sources of transaction costs provide four areas of action to observe in supply networks. Via searching, bargaining, monitoring and enforcing, social structure can manifest itself in other ways of performing actions necessary for transactions. In this way sources of transactions costs are a way to link theories from the bucket of *social structure* to actual transactions in supply networks. Each of the four sources is present in at least one of the two empirical studies carried out for this research.

1.2 Research method

This book uses gaming simulation as the research method. Using gaming simulation, researchers can study the behaviour of real people in a simulated environment (Roelofs, 2000). Gaming simulations are commonly used as training or learning tools (Wenzler, 2003; Druckmann, 1996; Gosen and Washbush, 2004; a.o.). Only limited use is made for research purposes, often positioned in the design sciences (Klabbers, 2006). This means that the research aims to test the effect of using a gaming simulation to train people or even to change organisations. In the analytical sciences, the sciences that empirically study a real-world phenomenon, few gaming simulations are used to collect data, most often of a qualitative nature to generate or test hypotheses. The methodological contribution of this book is the use of gaming simulations as a laboratory environment to generate and test hypotheses with both qualitative and quantitative data in the domain of supply networks.

Gaming simulation is a method in which human participants enact a role in a simulated environment (Duke and Geurts, 2004). Chapter 2 discusses gaming simulation as a research method. Section 2.1 describes that a gaming simulation consists of rules (what is allowed), roles (what a participant represents) objectives (how can you win, and what is positive or negative) and constraints (what are the limitations). This book answers the question as to whether gaming simulation can be a research method for (quantitatively) testing hypotheses in supply networks. It combines a repeatable experiment with the possibility to observe traders, transactions and the performance of a supply network. The repeatable experiment makes it possible to compare networks consisting of different people given the same controlled rules, roles, objectives and constraints. One can control for the 'load' (the value of configuration parameters of a gaming simulation) and the 'situation' (the way participants are selected and enter the gaming simulation, for instance as a part of a university course, obligatory or not, etc.), enabling testing of different parameter settings. Gaming simulation can easily be combined with questionnaires for the participants to gather data that cannot be observed.

Section 2.2 discusses the characteristics of gaming simulation to identify what this method can add for the analysis of the research question. Section 2.3 discusses functions of gaming simulation for non-research purposes, while Section 2.4 discusses the functions of gaming simulation for research.

In the domain of supply networks four research methods are widely used, i.e. questionnaires, case studies, action research and (computer) simulation. None of the four allows for the study of all aspects of the research question. In Section 2.5 an overview of the strengths and weaknesses of the four methods is given. Summarising here, the influence of social structure on the organisation of transactions can be studied in a single or small set of supply networks using case studies, to provide in-depth observations of actions and the surrounding context. However, the generalisability of detailed case studies is a complicated matter. Also it is hard to observe actual actions in case study research. Questionnaire research overcomes the generalisability issue, though lacks the in-depth knowledge of a subject within its context. Surveys do not observe actual actions either. Action research has similar disadvantages as case studies. Using computer simulation alone does not fit the explorative character of the research question. This book tries to overcome the disadvantages of the four methods mentioned by using gaming simulation, providing a proof-of-principle on how to answer a research question in the contextual-dependent domain of supply networks.

1.3 Two studies

Chapter 3 presents the research method as used in the two studies which form the core of this book. The research process is divided into two cycles (the design cycle and empirical cycle), facilitating the generation of hypotheses and testing them. The process is supported by two other cycles, being a multi-agent design cycle and multi-agent simulation cycle. Chapter 4 discusses the literature that is common to both studies. Theories on supply chains and networks are in the first section. The second discusses theories from new institutional economics and presents a framework by Williamson (2000) that is central to linking theories at different levels of analysis.

Two custom gaming simulations are conducted to study the influence of social structure on the mode of organisation of transactions. Chapter 5 presents the Trust and Tracing Game (TTG). The TTG directly assesses the influence of trust and embeddedness on the choice between the network and market mode of organisation. The TTG is a paper-based gaming simulation of a supply network of a product with a hidden quality attribute. Participants face the dilemma of whether to rely on trust or tracing when confronted with a possible cheat. Section 5.1 presents the results of the design cycle. It presents qualitative outcomes of 15 sessions with participants from both educational and business backgrounds. The section results in a list of variables to collect in future sessions to better understand the course of the action in the sessions. Section 5.2 uses this list of variables as input to start the empirical cycle. For this cycle, 27 additional sessions were conducted and analyzed. Social structure clearly manifested itself in trust and embeddedness influencing the organisation of transactions.

Resources are needed to conduct sessions with a gaming simulation. Participants can be hard to find, and should be willing to spend time playing. Especially when the combination of participants with a specific background is required, they should be considered as a scarce resource. Multi-agent simulation is a computer simulation method that models every person (an agent) as a distinct piece of software. Multi-agent simulations have multiple agent models running in one simulation to study emergent outcomes. Multi-agent simulation of sessions may provide a way to select interesting variable settings to test with human participants. The multi-agent model can be based on theoretical models or heuristically derived from observations of game sessions. Section 5.3 presents a proof of concept of multi-agent simulation of sessions with a gaming simulation. For the Trust and Tracing Game a multi-agent model has been developed, tested and validated. It has been possible to validate the multi-agent simulation on an aggregate level against sessions with human participants. Hypotheses were formulated based upon observations of sessions. Each hypothesis could be confirmed in the model runs.

The second gaming simulation, called the Mango Chain Game (MCG), is presented in Chapter 6. It was developed to study the bargaining power and revenue distribution among traders in the Costa Rican mango export network. The MCG assesses what factors, including social structure, determined the bargaining power, what mode of organisation was used and how this influenced the revenue distribution between traders. The data collection combined data from a questionnaire among the participants with the actual behaviour in the game session. Five sessions were conducted with smallholders in the Costa Rica lowlands, resulting in 82 contracts.

Chapter 7 provides the discussion and conclusions based on the findings in the two studies and elaborates on the practical experiences with gaming simulation as a research method. The first section goes deeper into the methodological conclusions and implications with special attention for the validity and reliability. Section 7.2 combines the conclusions of the TTG and MCG and links them to the main research question. The chapter ends with a number of implications for the domain of (food) supply chains and networks and possible applications in other domains.

2. Gaming simulation as a research method

Gaming simulations have been around for as long as humans. All children enact various social roles in play. Our species has even been termed *Homo ludens* (Huizinga, 1971). To use gaming simulation as a research method is a different kind of thing, and of a more recent date.

Gaming simulation is used in this book as a method to generate hypotheses and test them quantitatively using data derived from sessions. To be able to discuss how this works the first section introduces what gaming simulation is and of which elements it consists. Next, Section 2.2 discusses why gaming simulation can be used, what its inputs are and what its outputs.

Section 2.4 discusses the functions of gaming simulation for research. Hypothesis generation is linked to the more common approaches in the world of gaming simulation. Hypothesis testing is less common. Lastly this section discusses the use of multi-agent simulation for research with gaming simulations.

In the world of supply networks other research methods are more common than the one used here. Section 2.5 lists the four most important research methods (case studies, action research, questionnaires and computer simulation) and discusses their strengths and weaknesses. The position of gaming simulation compared to the other methods is discussed. The last section handles the issues of validity and reliability of gaming simulations.

2.1 What is gaming simulation?

The terminology in the literature on gaming simulations is ambiguous about when to use the term simulation, game, simulation game, gaming simulation (Duke and Geurts, 2004 use the term gaming/simulation). All concepts share the notion that human participants play a role in an artificial setting that models (an aspect of) reality.

Duke and Geurts (2004) discuss the definition, differences and combinations of gaming and simulation (pp 35-41) to position their 'policy exercise' as a particular type of gaming simulation. This book uses their positioning for two reasons: firstly, they are particularly shaped for the gaming simulation method that is used in this book. Secondly, Duke and Geurts are among the most theorising and structured users of gaming simulation in the field.

Duke (1980) defined simulation as 'a conscious endeavour to reproduce the central characteristics of a system in order to understand, experiment with and/or predict the behaviour of that system.' To be able to simulate one needs a model. Duke and Geurts use the definition of Leo Apostel (1960: 160), who defined a model as follows: 'Any person using a system "A" that is neither directly nor indirectly interacting with a system "B", in order to obtain information about system "B", is using "A" as a model for "B". The definitions used do not mention a structure, technique or implementation of a simulation or model. For the field of gaming simulation this is useful as the design may have many forms and incorporate many techniques, including complete sub-simulations to support the human play. According to Duke and Geurts (2004), a gaming simulation is *a special type of model that uses gaming techniques to model and simulate a system. A gaming simulation is an operating model of a real-life system in which actors in roles partially recreate the behaviour of the system*. The partial recreation by actors in roles could be completed with additional elements, ranging from simple forms or fiches to complicated computer simulations. In the current book the definition by Duke and Geurts will be used as the working definition of gaming simulation.

In previous research, Meijer and Hofstede (2003c) use the same elements as above in their definition, though their formulation is different. 'Simulation can be defined as studying the effects of human behaviour or decisions on some key variables in a model that represents a real-world system. A game can be defined as a clearly delineated activity with its own roles, rules and incentives, carried out for its own sake. A gaming simulation is a simulation of a real-world system in the form of a game. This implies that the roles, rules and incentives of the game mimic some real-world phenomenon' (Meijer and Hofstede, 2003c).

Wenzler is a gaming simulation practitioner coming from the world of large-scale business simulations. In Wenzler's nomenclature diagram all activities are called simulations. This is to emphasise the need for similarity with real-world systems. The realistic simulations Wenzler (2003) describes are used in training for new work practices or evaluation of new designs. More abstract simulations tend not to illustrate specific work situations but general principles. Wenzler calls these gaming simulations. To make the link with the definition of Duke and Geurts the more detailed simulations of Wenzler do not use gaming techniques but real-world rules, roles and objectives and constraints.

Around 2005 a new term came up that rapidly gained popularity: serious gaming. The term caught on in the computer (video) game industry mainly. World-wide the play of computer games is very popular and the (video) game industry is a billion-dollar-turnover sector. The transition from video games to serious games is to build games upon models that have a scientific background instead of fantasy environments. It is hard to pinpoint exactly the differences between serious gaming and gaming simulation as the terms are used interchangeably. Both simulate an aspect of the real world and both use gaming techniques. Often the role of the participant in a serious game is a fictitious one in which he or she needs to find information or explore the model underlying the game. This role is not part of the scientific model. A role in a gaming simulation is an essential part of the model. The participant plays a role and thus influences the run of the model. In practice the distinction between serious games and gaming simulations is vague. In this book, because of the aim to study the behaviour of participants, the term gaming simulation is used.

2.1.1 Elements of a gaming simulation and a session

To play a session with a gaming simulation one needs a design and a configuration. The design defines the institutional environment that spans the world within a session. The institutional environment is defined in 'rules', 'roles', 'objectives' and 'constraints'. These terms are based upon Gibbs (1974) to define the structure of gaming simulation. The configuration determines the values of the parameters of the design and the situational factors. In the present book, the parameter settings are called the 'load' and the situational factors the 'situation' of the gaming simulation.

The **roles** in a gaming simulation are divided into roles for participants and roles for game leaders. Roles do not necessarily match with the roles present in the real-world aspect modelled. Roles can be abstract interpretations of real-world roles or envisioned future roles that do not exist at present. Roles in a gaming simulation are distributed over the participants. Roles can differ in the actions that they can perform or in the objectives associated, or both. Game leaders can fulfil a role too, enabling a pre-scripted behaviour for a role. The game leader is like an actor in such a situation. When the role of a participant has nothing to do with the model itself (i.e. a fantasy role) I do not call the design a gaming simulation but classify it under serious games.

The **rules** in a gaming simulation are specific for a role or generic for all participants in the gaming simulation session. Rules can mimic behavioural limitations in the real world or can be artificial constructs to change behaviour of participants.

Objectives that participants experience in a session will most often be expressed in points or money to be earned. Different roles can have different objectives to create a multiple attribute incentive system. Objectives can be individual or the combined goal of a set of roles. Objectives are needed to steer the actions in a session. The gaming element of a gaming simulation means that participants will be motivated to win or do the best they can in a session. In some gaming simulations, especially for learning, the objectives of the individual roles are not aligned with what the game leaders are interested in. The incentive structure can be shaped to promote selfish behaviour while seen from an overall view cooperative behaviour would be better. The confrontation of participants with their behaviour compared to the optimal situation is used for learning.

Constraints limit the range of actions possible in a gaming simulation. They differ from rules as constraints limit the possible values of variables in a session. Rules shape the behaviour as they define what is allowed or forbidden. Constraints shape the values of punishments, the minimum and maximum values of time, money and points. Constraints can be set to make a gaming simulation match the real world or to make it explicitly different.

The roles, rules, objectives and constraints are part of the design of a gaming simulation. A design itself often allows multiple ways to tune and use a gaming simulation. To be able to actually conduct a session two more inputs are required, namely the load and the situation. The load and the situation are not a part of the gaming simulation.

The **load** can be defined as the values of the variables in the design of the gaming simulation. The term is used after Wenzler (2003). In the world of systems engineering it could be called the 'state', and in the world of multi-agent systems the 'variable setting'. Gibbs (1974) uses the term 'scenario', but with the current use of this word by the computer game industry the term has a changed connotation. Without alterations to the design a gaming simulation can be shaped for different configurations. How many participants in a certain role, how much money or how many points are in a session at start-up, what is the value of punishments, et cetera? Depending on the complexity of a design, loads can lead to completely different outcomes of sessions. This gives the opportunity to use the load as independent or mediating variables in otherwise fully identical gaming simulations. Experiments can be set up using variations in the load, or with a constant load and variations in the situation.

The **situation** of a session is all variables that surround a session but are not part of the design. The venue, selection of participants, the reasons for and consequences of participation, etc, are all part of the situation. Many of the situation parameters are determined through the actual organisation of a session. One of the most important variables in the situation when using gaming simulation for (quantitative) research is the ability to vary the selection and preparation of the participants. What do you tell them in advance? What is the social background of the group (culture, homogeneity, do they know each other), et cetera?

2.2 Reasons to use gaming simulation

A gaming simulation makes people operate in a model, thus eliminating the need to build in psychological and sociological assumptions (Duke and Geurts, 2004). The actions of players within a gaming simulation consist of a set of activities aimed at achieving goals in a limiting, sometimes abstract context with many constraints. As participants take values and beliefs from their real life with them into a session, e.g. culture, it can be made part of a model, without the need to formalise it in a (computer) model.

In the domain of supply networks (and perhaps even in the social sciences) research methods either study a phenomenon within its context with its associated issue of generalisability, or context-free to build a universal theory with the associated risk of missing the influence of contextual factors. This dichotomy in approaches boils down to the question as to whether it is better to study a phenomenon in depth, providing all possible links with contextual variables and thus losing the general applicability of the conclusions, or to study a phenomenon in a broad sense, using many respondents / subjects from various backgrounds and drawing a conclusion that is more universally true, but may have missed important contextual issues.

Supply networks are known to be context dependent, for instance because of the product characteristics or geographic location. No wonder a research method that incorporates contextual variables like the case study method is popular among network researchers.

Gaming simulation as a research method provides an intermediate step between the study of a case in a real-world context and the more context-free methods like questionnaires. A gaming simulation emulates a controlled context and thus provides a repeatable experiment. Participants can be selected from real-world supply networks or from other populations. Large numbers of participants, using participants with different backgrounds or from different networks and the possibility to manipulate the context in a controlled way make gaming simulation a laboratory-style experiment in terms of controllability with a context that resembles the real-life situation, as in case studies.

Figure 2.1 shows the inputs and outputs of a session with a gaming simulation. The six inputs just described determine the full environment present in a session. The only input that cannot be controlled is the group of participants. One can select participants and one can instruct them about the rules, roles, objectives and constraints but that cannot take away the relational history, experiences and world view of the humans that participate. On the contrary: because the six inputs on the left of Figure 2.1 are controlled, the participants are the only uncontrolled source of variation which makes a gaming simulation an ideal tool to test behavioural responses in a controlled environment. The session will yield qualitative and quantitative data. The participants gain an experience. This experience could be debriefed and made valuable for the participants themselves. The application of gaming simulation for learning purposes is based upon this experience. For research purposes it is important to note that participants with the experience of a session are different from naïve participants.

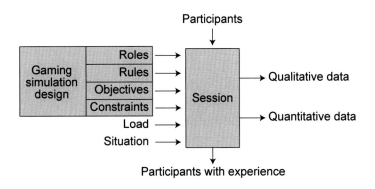

Figure 2.1. Inputs and outputs of a gaming simulation session.

2.3 Gaming simulation for non-research purposes

While in this book the focus is on the application of gaming simulation for the various functions in research, gaming simulation can be used for other functions too. These functions are mainly teaching and training. In the research conducted here there is a learning experience for the participants in a session. The learning experience was the primary motivation for participants to engage in a session. This avoided the need to pay participants as is sometimes the case in economic experiments.

There is a long history of teaching and training with gaming simulations. Although the current increase in attention for serious gaming might give a different impression, gaming simulation is certainly not new. Our species has even been termed Homo Ludens (Huizinga, 1971). No wonder, then, that gaming simulations have become, since the nineteen sixties, a successful training tool for many areas of organised life (Crookall, 1994). Military strategies have been developed with gaming for centuries (Stoop, 2008). In the domain of chains and networks the Beergame (MIT, 2008) is famous and played all over the world.

2.3.1 Learning

Positive learning effects of playing for young children are widely accepted in educational studies. The leading opinion is that children practice their social skills while playing. Gaming simulations (and serious games) are often used for learning practices. The range of complexity of the designs differs tremendously. From extremely simple abstract games by game-guru Thiagi (Dr. Sivasailam Thiagarajan) using a paper form and a good game leader to multi-day-multi-player management games like the Global Supply Chain Game (Corsi *et al.*, 2006) and gaming simulations by consultancy firms (Wenzler, 2003) all are meant for practising a role and learning new insights through experimentation and feedback. Gosen and Washbush (2004) studied the effectiveness of gaming simulations for learning. Summarising the best 39 papers out of 115 papers that describe results of gaming simulations, they conclude that at face value the effectiveness of both computer-supported gaming simulations and experiential exercises (more simple paper-based gaming simulations) is clearly supported by these papers. They do criticise, however, the lack of a shared measurement methodology to study the effectiveness. They even go so far as to conclude that: '...the criterion variable being used, which is learning (...), is illusive. To our knowledge, every attempt to concretise this variable has failed. As a field, we believe we know what it is, but what it looks like so it can be measured lacks form. (...) The problem lies at least in part in the nature of learning and the learners' expressions of it. Learning is an internal mental process and what is learned and how it is learned is unique for each individual. To create an instrument able to capture what is actually learned is easier said than done. To get at this, learners have to be motivated to express their learning,(...)'. Most studies Gosen and Washbush (*ibid.*) found used self-report of participants to indicate the learning. The second-largest group used objective measures. Druckmann (1994) investigated the effectiveness question too, and suggests that simulations with more 'fidelity' (similarity to a real-world situation) are better suited for learning actual practices, while more abstract simulations are better for learning insights. This is in line with the reasoning of Wenzler (2003). The usefulness for learning insights in complex problems has been most prominent (Druckmann, 1994).

2.3.2 Design-in-the-small and the policy exercise

Klabbers (2008) presents a meta-framework that tries to span a theoretical space that embodies all types of gaming (simulation). It is the result of his many years writing articles about and performing gaming simulations on complex social systems. Klabbers distinguishes between two types of design. Design-in-the-large is about changes made in social systems that exist in the real world. His work as a consultant has brought Klabbers into many organisations and governmental issues where changes in a social system had to be designed. Therefore he used gaming simulations that were a design-in-the-small of the real-world situation that should be 'designed-in the-large'. In a 'design-in-the-small', solutions, future situations or problems can be enacted and analyzed, leading to a hypothesised possible solution for the design-in-the-large. Here the link can be made between Duke and Geurts (2004) and Klabbers (*ibid.*) focus on the analysis and exploration of current and future social systems. De Caluwé and Geurts (1999) analysed the use and effectiveness of gaming simulation for strategic culture change and concluded the approach was both useful and effective within certain limitations.

Duke and Geurts (2004) mention five functions of gaming simulation that make them useful for their policy exercise (which is a special type of gaming simulation). They summarise the functions as the five 'C's: Complexity, Creativity, Communication, Consensus and Commitment to action. The complexity function is about the way in which gaming simulations can help in analysing complex situations with multiple actors and ill-defined problems. Creativity is about the way that gaming simulation can help people think outside of their normal environment. The last 3 C's help change a real-world (social) system into the one that has been designed in a gaming simulation (Design-in-the-Large from Design-in-the-Small to quote Klabbers). Communication between participants is facilitated by having them in the same room talking about the same topic, Consensus is created by the shared experience and shared formulation of the problem and possible solutions, and Commitment to action is gained from the possibility to successfully enact a future state.

2.4 Gaming simulation for research purposes

Gaming simulation is a method with many faces. The learning function is most prominent and the majority of published papers using gaming simulation are about learning and training applications. In this book the emphasis is put on the functions that are important for research. Three major functions are described below: hypothesis generation, hypothesis testing and multi-agent modelling.

2.4.1 Gaming simulation for hypothesis generation

Grounded theory is theory that is 'derived from data, systematically gathered and analyzed through the research process. In this method, data collection, analysis, and eventually theory stand in close relationship to one another. A researcher does not begin a project with a preconceived theory in mind (...). Rather, the researcher begins with an area of study and allows the theory to emerge from the data. ... Grounded theories, because they are drawn from data, are likely to offer insight, enhance understanding, and provide meaningful guide to action' (Straus and Corbin, 1998).

The grounded theory approach first starts with *description*, the use of words to convey a mental image of an event, a piece of scenery, a scene, an experience, an emotion, or a sensation; the account related from the perspective of the person doing the depicting (Straus and Corbin, 1998). A description differs from theory in that it is not a set of developed concepts that are related through statements of relationships. The description is just the framework, through which theorising can be done. The second step is *conceptual ordering* of data into discrete categories according to their properties and dimensions and then using description to elucidate those categories. The last step is *theorising* which is the 'act of constructing (...) from data an explanatory scheme that systematically integrates various concepts through statements of relationship' (Strauss and Corbin, 1998). The explanatory scheme is not yet empirically tested in a quantitative way but has qualitative data as evidence. It can be used as a hypothesis for quantitative research. In this book it is called an 'induced hypothesis'.

Within the world of gaming simulation a qualitative search for an explanatory scheme is a common application of the method. Duke and Geurts (2004) developed a 21-step sequence for the development and implementation of policy exercises: a special type of gaming simulation designed together with the stakeholders. Since Duke's first book in 1974 their work has been focussed on the scientific approach of gaming simulation. In their book, Duke and Geurts (*ibid.*) work out each of the 21 steps and give hints and tips for best practices. When discussing their methodology they put the emphasis on the contribution of theories and modelling practices from bodies of theory like sociology and social psychology. They present gaming simulation as a profession. While their efforts in formalising the gaming simulation process are valuable, the problem arises that their methodology is not easily comparable with other methodologies. The way in which Duke and Geurts (*ibid.*) operationalise the model before starting observations is very similar to the Grounded Theory approach (Strauss and Corbin, 1998). When phase I and II of Duke and Geurts' approach is considered outside the scope of a client with a problem situation, the actions taken are very similar to the grounded theory approach. Step 2 could be called *description* and steps 6 and 7 could be called *conceptual ordering*. Step 14 and the actual play with the client are close to *theorising* as they test or try to predict a future state of the problem situation. Table 2.1 gives an overview of the link between Grounded Theory and Duke and Geurts (*ibid.*), and lists this book in the third column, which is described in Chapter 3.

Table 2.1. Relationship between Grounded Theory, Duke and Geurts (2004) and this book.

Grounded Theory (Strauss and Corbin, 1998)	Duke and Geurts (2004)	This book (see Chapter 3)
Description	Phase I; step 2: define the macro problem	Unstructured session transcript (#9)
Conceptual ordering	Phase II; step 6 + 7: defining the systems – content, boundaries, interrelationships / displaying the system – create a lucid cognitive map	Finding structure in unstructured session transcript in next iteration of the design cycle.
Theorising	Phase IV; step 14: build, test and modify a prototype exercise + Phase V; step 18: facilitating the exercise	Induced hypotheses (#10)

2.4.2 Gaming simulation for hypothesis testing

While hypothesis generation is common using gaming simulation, the (quantitative) testing of a hypothesis is less prominent. This section explains why using gaming simulation for this function can be fruitful.

In the world of science, paradigms frequently act as walls between disciplines. Paradigms assemble communities of researchers around a set of problem situations that these people have defined as relevant, and a set of methods that they have defined as acceptable (Kuhn, 1962). In the world of gaming simulation, the Journal *Simulation & Gaming (S&G)*, ISAGA (International Simulation and Gaming Association) and the entertainment game industry set the standard. The majority of the published research papers from these three favour the science of design.

Hofstede and Meijer (2008) argue that there is no methodological reason why gaming simulators should not employ some of the methods of empirical sciences when appropriate. Yet the potential role of empirical science is a controversial issue around gaming simulation. Apart from employing different techniques, the 'two worlds' employ different concepts of causality. Within the design sciences the major scheme is: build and evaluate in context-of-use, using process causality (Klabbers, 2006). Within the analytical sciences tradition the major scheme is: develop a theory and test or justify it, using variables and correlations. A corollary caused by the need for statistical significance is the tendency to rule out as many context variables as possible.

The two types of causality do not need to bite each other. In the world of information systems this is more common (Benbasat and Zmud, 1999). Checkland and Scholes' (1991) action research is another long established example. Using a survey or other empirical technique for getting to know about a real-world problem situation is not controversial. But using a set of implementations of a design as a source of data outside the original context of the real world without the design can be controversial because action researchers doubt the relevance of these data, and perhaps with good cause. This is due to the empirical researcher's tendency to discard context. Hofstede and Meijer's point is that *discarding context is not a prerequisite for doing empirical research*. The following paragraphs focus on how to integrate methods in gaming simulation.

Klabbers (2006) presents a meta-framework that tries to span a theoretical space that embodies all types of gaming (simulation). It is the result of his many years writing articles about and performing gaming simulations on complex social systems. Klabbers distinguishes between two types of design. Design-in-the-large is about changes made in social systems that exist in the real world. In his words, the real-world situation should be 'designed-in the-large'. In a 'design-in-the-small', solutions, future situations or problems can be enacted and analyzed, leading to a hypothesised possible solution for the design-in-the-large. Here the link can be made between Duke and Geurts (2004) and Klabbers.

Klabbers (2008) discusses the evaluation method of gaming simulations and pays particular attention to the question as to whether a gaming simulation can be studied using an analytical science approach. He comes to the conclusion that gaming simulations are 'non-trivial machines' that are, by definition, characterised as *synthetically indeterministic (....), history dependent, analytically indeterminable* and *analytically unpredictable*. By non-trivial machines, Klabbers refers to a system view in which the relationship between input and output cannot be determined because the system changes itself due to history and possibly other mechanisms that are part of the system.

Klabbers reasons that because of the internal feedback mechanisms in a non-trivial machine, it is not possible to study gaming simulations with a methodology used in the analytical sciences domain. His reasoning is particularly shaped towards the evaluation of the 'artefact' that a gaming simulation is for learning or a change in an existing situation. In the current book the aim is not to assess the effect of participation on the participants or on any existing situation. The primary purpose is to gather data about the behaviour of participants in a (simulated) context. In Figure 2.2, Klabbers' approach would be the evaluation of the vertical flow of participants through a session. The current book studies the horizontal plane of independent and dependent variables. The participants should be described thoroughly (on the relevant aspects) to be able to use their history and backgrounds as an independent variable. Their behaviour in a controlled context then is a trivial (though likely to be non-linear) machine. Klabbers briefly mentions the possibility of the approach used in the current book (Klabbers, 2008: 188) when he says that: 'most games – artefacts – can be used as tools

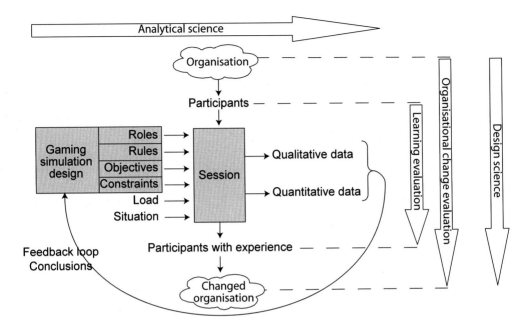

Figure 2.2. Inputs and outputs of a session related to the design and analytical science.

to test simultaneously various domain specific theories... Each specific theory will enlighten certain characteristics of the game-artefact, while ignoring other ones'. I would argue that the theory to be tested shifts the focus to applicable variables and mechanisms of study, but not necessarily to characteristics of the gaming simulation itself. It is the challenge to uphold the principle of control, randomisation and comparison that Klabbers (2008: 188) gives in case one really wants to use analytical methods with gaming simulation.

The previously mentioned conclusion of Hofstede and Meijer (2008) that little attempts have been made to mix design science, analytical science in simulateous interaction with the real world in the way they propose with gaming simulation could be due to obvious practical reasons; most gaming simulations are just not repeated often enough, or if so, results are not kept in comparable format. But there are also paradigmatic reasons, in that methods are deemed incommensurable. Hofstede and Meijer believe this to be falsifiable, and give examples from their practice, among which is the Trust and Tracing Game that is part of the current book too. Klabbers (2008) gives three illustrations of integration between artefact design and analytical science using gaming related artefacts. Noy *et al.* (2006) used two gaming simulations and a computer simulation to test several social theories. Their study is '*a clear-cut example of the analytical science approach, applying the scientific method and the variance theory of causality. From a theoretical perspective, one should understand that the theories to be tested are still disconnected, and through the use of three different artefacts – based on distinct minitheories – they cannot be integrated in one nomological network. Nevertheless, Noy et al. (2006)*

offered two frames for linking the artefacts and the theoretical concepts. A fully integrated theory can only be achieved when only one game (gaming simulation, SM) *is being used, specifically designed to incorporate the relevant theories (...). Through changing the parameters of that game, the independent variables for the experiment, various dependent variables (...) can be measured and causal links can be tested. Such a specifically designed game would present a pure case of theory development and testing. The added value of such an effort will be that form, content and meaning of information and knowledge are uniquely related to one another. The end result will be a more robust theory.'* (Klabbers, 2008: 190) The second example of Klabbers does not use a gaming simulation but a multi-agent simulation and is therefore discussed in the next section. The last example he gives are Mallon and Webb (2006), who used an analytical approach for games analysis. Their methodology was a phenomenological one, and bottom-up and inductive. Within the framework used in this book it holds the middle ground between the hypothesis generation and hypothesis testing as the research first followed a grounded-theory-style approach. The data they used were not generated in the session itself but during a session in testing the response and emotions of participants, not using the actual data from actions within a session. It is also true that this study was about the analysis of computer games, not of a social dilemma existing in the real world. They used an analytical approach to study the quality of a design, meanwhile developing a series of evaluation criteria that could be added to evaluation methodologies.

In a strict sense the use of gaming simulation to explore complex problems (like the Duke and Geurts (2004)-approach; Mayer, 2008) can be seen as a qualitative use of gaming simulation in the analytical sciences. In exploring and formulating complex problems (macro problems as Duke and Geurts say) the function of the gaming simulation is to gather qualitative data from the play of the participants in a gaming simulation that is not yet a fully enveloped model about a good future state of the real-world that is being modelled. From the qualitative data a hypothesis can be formulated in the form of an envisioned future state of the system. New sessions to be played with the model of a future state to make participants experienced or willing to change or other attempts to actual change attitudes or behaviour should be placed in the design sciences, as the function of the gaming simulation is to change participants behaviour or attitudes and via this an organisation. When the aim is to change organisations via the participants the link with Klabbers' concept 'design-in-the-small' is made.

Gaming simulation can be seen as an alternative approach to (computer) simulation. A computer simulation of human behaviour in supply networks would model the supply network and model humans with for instance decision rules, and then study the behaviour that emerges in model runs. A gaming simulation models the supply network, but does not model humans, but integrates them into the simulation by giving them a role. The behaviour can be studied in sessions.

The use of gaming simulation as a hypothesis testing tool is a logical extension to the method of Duke and Geurts (2004) for complex policy problems, as they emphasise the importance of operationalisation of the key concepts used in a gaming simulation. Operational (and measurable) concepts are therefore required. In the field of policy research there are ample examples of studies using a gaming simulation to test new procedures. Some are close to role plays using actors as stimuli (Van Laere *et al.*, 2000). Most of the studies use qualitative observations to determine differences between groups or treatments. Pre- and post questionnaires and the debriefing or discussion afterwards are the most common sources of data (Bekebrede and Mayer, 2006, Corsi *et al.*, 2006; a.o.).

Few authors used the data generated in the session itself. Roelofs (2000) tested a mapping technique for structuring policy issues using a gaming simulation as test bed. The approach is similar to the current book as she explains behaviour within a session with data gathered in a session, but differs in the nature of the data (qualitative) and the domain (policy research). Methodologically Roelofs (2000) can be seen as the qualitative predecessor of this book. Kuit *et al.* (2005) used a computer-supported gaming simulation to investigate strategic behaviour in a deregulating energy market. Although their model is numeric their results and conclusions seems to be based on qualitative analysis of participants' behaviour. In their paper Kuit *et al.* mention results and findings that are backed up by observed actions, but they do not mention a quantitative analysis. De Caluwé (1997) conducted research after an organisational culture intervention using a gaming simulation as an intervention tool. The data he used were not collected from within the sessions but from interviews and questionnaires before and after the intervention. Gasnier (2008) studied patent management by triangulation of questionnaires and a gaming simulation. He used data from the sessions to tell what happened, but studied the effects of participation on learning and the real-world situation, similar to De Caluwé whose framework he also used.

Within the domain of supply chains and networks a few gaming simulation approaches come close to the methodology used here. The number of computer-based network simulations used for training is growing rapidly (Meijer and Hofstede, 2003a; Van Liere *et al.*, 2004). A computer-based gaming environment facilitates quantitative data collection about actions in a session. The author does not, however, know about any gaming simulation that tries to quantitatively explain why participants act as they do in a session using data from the session. In the world of (commercial) training the facilitators are able to give feedback on the actions taken in a session for explaining the outcomes. They can discuss this with the participants (Corsi *et al.*, 2006). Although the data seems to be available and the methods used are very similar, giving feedback is not the same as trying to test or even induce hypotheses about behaviour. Hofstede *et al.* (2003) developed a computer-based chain game for distributed trading and negotiation. This tool promised to facilitate data-gathering from actual sessions but has not been applied in actual research projects, due to lack of funds to finish the final stages of the software. A difference between the computer-based chain games and the methods employed in the current book is the absence of face-to-face negotiation in a computer environment that could be a major determinant for the emergence of real-world social relations in a session. Chat and e-mail conversations are different from face-to-face ones, and a whole body of theory about communication channels should be introduced to relate outcomes to face-to-face business.

2.4.3 Sensitivity analysis: multi-agent simulation of a gaming simulation

As has been said before, both gaming simulation and computer simulation are ways to simulate an aspect of a real-world or future system. There are many definitions of simulation, and the notion of simulation from the definition of Apostel (1960) as used before is an abstract one suited for gaming simulation and computer simulation. As Klabbers pointed out it is valuable to be able to formalise a design-in-the-small, either through incorporation of formal (computer) models in the design of a gaming simulation or through a separate multi-agent simulation. The reason to apply multi-agent simulation in the current book is that sessions require many new participants for each experiment. A multi-agent simulation could be a tool that in the long run could be used to select the most interesting experimental setups to play with human participants. The ability to perform sensitivity analysis on a validated model of trade behaviour in a game session could prevent researchers using the time and availability of participants in sessions that have variable settings not leading to interesting outcomes.

It falls outside the scope to discuss the ins and outs of multi-agent system design here as it is a separate discipline in itself. (See Wooldridge and Jennings (1995) and Castelfranchi (1998) as authoritative sources). Chapter 3 shows how the multi-agent simulation is linked to de design and empirical sessions with a gaming simulation.

In the case of multi-agent simulation of a social system the number of variables to incorporate would soon become enormous if the full context that could be of importance were to be incorporated. Therefore only a few selected elements of the context are operationalised in a multi-agent design. One cannot (yet) fully model a human with all its richness in social behaviour, reasoning and culture. An agent representing a human in a multi-agent simulation will always be a thin model of an aspect of a real human being. Integration of the gaming simulation approach with multi-agent simulation helps reduce the complexity of the environment. As Figure 2.1 showed, the rules, roles, objectives and constraints plus the load and situation are known inputs. The only 'uncontrolled' variable is the participants. The roles they play and the amount of actions are defined in the design of the gaming simulation, thus steering the aspects to be incorporated in the agent model.

The CIRAD institute uses combinations of participative development of multi-agent (computer) models and role-playing games (Barreteau, 2003 discusses their approach). They do not use raw data from the gaming simulation directly, but only the (qualitative) reactions of participants to help modelling the computer models.

Rouchier and Robin (2006) is the second example that Klabbers (2008: 190) gives for integrating analytical sciences. Robin studied the rational behaviour of individuals in a market environment by combining experimental economics with a multi-agent system. The data from experiments are fed back into the multi-agent system, and then the system helped to explain outcomes of the experiments. In the field of behavioural economics (concerned with individual actors' economic behaviour) this type of interplay between artefact and experiment is common. When studying larger socio-economic systems the aforementioned complexity of the context becomes more dominant and the use of a gaming simulation as an intermediate step between the real world and the object of study could be useful for multi-agent simulators. The current book is a proof of principle of this approach.

Coming from the multi-agent simulation approach, Mizuta and Yamagata (2005) present a gaming simulation method that comes close to the approach in the current book. They developed a multi-agent simulation of the international greenhouse gas (CO_2) emissions trading market and performed experiments with it. Additionally they developed a web-based application in which the simulated trading agents were replaced by real human people performing the trader role to create a gaming simulation. They let students play the gaming simulation and found from the data generated within the gaming simulation an emergent dominant trend and a learning effect among the participants. The approach of Mizuta and Yamagata differs from the one used here in the absence of face-to-face negotiation and social issues in trade, the lack of a supply network aspect as the CO_2 emissions market has no product to produce except for clearing the market, and a different approach of the hypothesis generation, testing and milti-agent simulation functions of gaming simulation. They started with a formalisation phase by building a multi-agent model upon theoretical models of Walrasian auctioneers and Double Auctions. Then they used a gaming simulation to test the formalised model with real human participants. Hypothesis generation did not take place with their gaming simulation.

2.5 Position of gaming simulation among research methods

In the domain of chain and network studies gaming simulation is not a common research method. Case studies, questionnaires and computer simulation are widely in use. Action research is less common, but still present in numerous papers. Case studies, questionnaires and action research are part of world of the analytical sciences. Computer simulation is part of the design sciences. This section compares the common research methods with gaming simulation to position the latter method in the domain. Table 2.2 gives some strengths and weaknesses for each of the five methods mentioned. This table is based upon readings on research methodology like Bryman and Bell (2003) and Creswell (2002), among others.

A case study of a supply network is a well-known research method in the CNS domain. Yin (2003) gives a two-part definition of a case study. The first half gives the scope of a case study:

> *A case study is an empirical inquiry that*
> - *Investigates a contemporary phenomenon within its real-life context, especially when*

Table 2.2. Strengths and weaknesses of research methods in common in supply chain and network studies.

	Case studies	Questionnaires / Surveys	Action research	Computer simulations	Gaming simulations
Strengths	Real world in-depth study. Observation of actual actions and direct communication.	The power of large numbers, wide range and number of respondents possible. Little disturbance of the actual behaviour. Well-known method, incl. Solving issues like non-response, etc.	Observations from within an (organisational) situation. Observation of the actual behaviour. Longitudinal observations with possibility to find patterns that will not be found using iterative observation moments.	Virtually unlimited numbers of experiments. Any possible setting can be tested. Testing hypothesised models with endless variation of the environmental and internal variables.	Repeatable experiment. Observation of actual actions and behaviour. Control over environment
Weaknesses	Low repeatability due to changing contexts. Generalisability complex due to contextual bindings	No control over environment. Little information about context. Answers can be socially acceptable instead of real behaviour.	Low repeatability due to changing contexts. Influence of researcher on process. Generalisability can be complex, due to observation of one situation within its context.	No real observations. 'Rich' human is modelled but can you model the tacit knowledge?	Simulated context, not for real. Large number of participants willing to spend time required.
Key references	Yin (2003)	Churchill (1999)	Checkland and Scholes (1991)		Duke and Geurts (2004), Klabbers (2008)

- *The boundaries between phenomenon and context are not clearly evident.*

Yin positions the case study method as a tool to cover contextual conditions in research problems where the context is believed to be highly pertinent to the phenomenon of study.

Case study research can have multiple forms. Single case studies investigate only one real-life case. Multiple cases can be investigated and compared using a comparative case method (e.g. Agranoff and Radin, 1991). The topic of research of this book could have been conducted using a comparative case study design to obtain the large number of observations of transactions, agents and networks required. The time typically required for each case would have been an obstacle for a PhD project. Secondly, a comparative case study approach would have included other context than the social one alone, making networks more difficult to compare and distracting from the research topic of this book.

The questionnaire method can deal with both phenomenon and context, but its ability to investigate the context is limited due to the constant struggle to limit the number of variables to be analyzed. It is not possible to observe either the actual behaviour or the actions of people, contrary to gaming simulation, case studies or action research.

Action research has the advantage of observing not only the actions in a network but the actual day-to-day behaviour too. A gaming simulation offers this possibility too, though in a simulated context instead of the real world. The generalisability of action research is problematic due to the context dependency, but its topics can be relevant as these are the actual problems in day-to-day operations in supply networks.

Computer simulation in the domain of supply chains and networks is common, especially when analysing logistics and production management. But as Omta *et al.* (2001) pointed out in their first editorial of the Journal on Chain and Network Sciences, research in the domain is integrated and includes managerial aspects of people and techniques. To simulate human behaviour is complicated in a domain where the contextual variables are important and problems are complex. The risk of missing essential elements of human behaviour is paramount.

2.6 Validity, reliability and repeatability of gaming simulation

Gaming simulation methodology is like developing a new research tool for each specific situation. Every new research tool needs to be calibrated, regardless of whether it is about chemistry or economics. The reliability of the tool in different situations needs to be tested. Is the tool robust enough to withstand disturbances from outside, like people walking into the venue, differences in room lay-out and other external influences? Are the rules, roles and incentives clear to all participants? Solutions for these questions fall into the 'craftsmanship' of the designer of gaming simulations and game leaders, but results can be tested. Kriz and Hense (2006) offer one of the more theory-based evaluation frameworks. One can only test a new design with test groups of participants. By asking their feedback on the design and systematic observation of the process better designs can be made. It will depend on the experience of the 'craftsmen' how fast a design will be found that fits the list of requirements. The question comes up as to what the influence is of the designer and game leaders on the outcomes of the sessions.

In their approach Duke and Geurts (2004) emphasise the importance of the thorough operationalisation of concepts into the gaming simulation model. A systematic check of the rules, roles and incentives against theoretical and stakeholder-constructed concepts is part of their 21-step process model. In the current book the essence of their approach is used and applied to models behind the TTG and MCG. The models are not the result of stakeholder input but of theoretical concepts. The question remains as to what the rationale is behind choosing card-board cups as a model for a stock. The answer regarding the rationale is two-fold. The first part of the answer is that following Duke and Geurts (*ibid.*), one needs to ensure that the model of the aspect system under study (from whatever source it may be: stakeholders or theory) can fully express itself in the gaming simulation. Here a systematic scientific approach towards the design and application comes into play. Klabbers' (2006) positioning of gaming simulation in the design sciences can help in the development of a gaming simulation. I refer to Klabbers, and Duke and Geurts as authoritative sources. The theory-based evaluation of design of gaming simulation by Kriz and Hense (2006) provides a way to evaluate a design.

Secondly, the facilitation of sessions is an art or at least an act of craftsmanship. Experienced facilitators have certain sensitivity towards solutions, materials and the way in which rules, roles and incentives are operationalised. One can script each and every part of this process, and try to control for every variable, including the colour on the walls of the venue, the background music in the canteen and the way noisy participants are dealt with. Experienced facilitators have a basic understanding to what extent these circumstances and process aspects influence the outcomes of a session. In test sessions the facilitation can be tested. In the book 'Why do games work? In search of the active substance' (De Caluwe *et al.,* 2008) several authors reflect upon the facilitation. Here we refer to that book as a state-of-the-art discussion on facilitation.

2.6.1 Validity and reliability

The validity of a gaming simulation is a point of discussion among fellow gamers. As Druckmann (1994) pointed out, there are the believers and the non-believers and too often the discussion between the two groups is impossible due to the non-rational reasons. The most common critique for behaviour observed in a session is 'it is only a game...'. In the light of the methodology used in the current book this discussion deserves attention.

In the literal meaning the statement 'it is only a game' is true. A gaming simulation is a model of reality, and the roles, rules, objectives and constraints are necessarily different from the real world. The insinuation of the statement is, however, that behaviour observed in a session is unlike behaviour in the real world and is no valid representation of real-world behaviour. Peters *et al.* (1998) discuss the validity of games (gaming simulation) based upon the work of Raser (1969) who defined validity of models in the following way: 'A model can be said to be valid to the extent that investigation of that model provides the same outcomes as would investigation in the reference system.' Raser (1969) suggests four aspects of validity that apply to gaming simulation:

- Psychological reality → To what degree does the gaming simulation provide an environment that seems realistic to the participants?
- Structural validity → To what degree is the structure of the gaming simulation (the theory and assumptions on which it is built) isomorphic to that of the reference system?
- Process validity → To what degree are the processes observed in the gaming simulation isomorphic to those observed in the reference system?
- Predictive validity → To what degree can the gaming simulation produce outcomes of the historical or future reference system?

The psychological reality demands that sessions are conducted in such a way that participants are emotionally involved and really play their role. The situation of the session in the life of the participants, the consequences of participation or non-participation and the location and atmosphere of a session and its moderation are important factors. This requires craftsmanship of the game leader that is hard to operationalise in a scientific context. Various authors have made attempts at determining the quality of conducting sessions. Kriz and Hense (2006) offer an elaborate and theory-based evaluation methodology, that according to Klabbers (2008) does a good job in (temporarily) bridging the gap between analytical and design sciences. Kriz and Hense's approach is an adapted version of the theory-based evaluation method by Reynolds (1998). They distinguish between concept, design and application that can be evaluated. The application is similar to conducting the sessions in this book. The concept and design of Kriz and Hense are grouped under 'design' in the current book.

In relation to the structural and process validity, Kriz and Hense first developed a logic model based on theory of the gaming simulation under evaluation. They distinguish between the logic model that is built into the simulation itself (the aspect model to be simulated) and a logic model of the actual run of the gaming simulation in order to facilitate learning, for example. Secondly they measure the effects of participation in regard to the logic model's outcomes. In a third step, data is collected on mediating and background factors. They are also derived from the logic model. The fourth step consists of estimating the main effects of participation and the fifth of testing causal mechanisms of the logic model to explain these effects. '*Here is where the theoretical approach can show one of its most important merits: additionally to learning on the effects of the game, the logic model can be used to analyze which factors contributed the most, and which factors had a detrimental effect. In a sixth step Reynold's (1998) approach proposes to interpret the findings for the purposes of generalisation and knowledge transfer*' (Kriz and Hense, 2006). Their final step actually contributes to improvement. Here, the results of the above steps are used for identifying possible areas of improvement of the gaming simulation or its implementation. Guided by the logic model and analysis of its causal implications, weak spots may be found as well as particularly important factors for success.

The structural and process validity of a gaming simulation can be tested during the design phase. Through repeated sessions with similar participants using varying loads and situations the sensitivity can be explored. Duke and Geurts (2004) call this the 'rule of ten' (sessions) that need to be played before a gaming simulation is ready for in-situ use.

Testing for the predictive validity using gaming simulation is complex. In the current book all gaming simulations are issue-based, instead of simulated environments in which anything is possible. One could vary the load and situation in such a way that it closely mimics a real-world situation. Playing a session should show the same issues that occur in the real world. The predictive validity of gaming simulations that create a virtual world in which many issues can emerge from all possible actions is hard to test but falls outside the scope of this book.

In trade games like the ones in the current book the question is what makes people act like they do in a session. In the real world a trader is supposed to act according to the (economic) incentives perceived moderated by the social norms and personal beliefs the person holds. In the real world the incentives will be more complex than in a session. Multiple issues will ask time and attention of a trader, and most likely there are a number of deals possible and even necessary to run a smooth business. In a gaming simulation the rules, roles, objectives and constraints will be scaled down to the essential elements of focus for the model to operationalise. This requires an optimal balance between the four criteria for validity. While for instance a rule about a certain sequence of steps may improve the process validity, it may take the participants out of the flow of the session when they have to wait often, reducing psychological reality.

Neither the personality of a person is changed in a session, nor is his culture. The way personality and culture express themselves in actions may be different from the real world as the consequences of one's action are different. If we use cheating as an example, it might be accepted in a culture to make fun of each other by cheating upon each other when the gaming simulation allows doing so. As long as the cheats do not really harm the person that is cheated the participants will laugh about it in the debriefing. The structure validity in this example depends on the culture participants come from. The incentive to cheat is changed as the consequences are small compared to the real world. In sessions with people from a culture where cheating is taboo, the participants are not likely to cheat at all. The message here is that as the incentive system is different in a gaming simulation the expression of culture and personality may be different, but the behaviour of the participants still follows the incentive structure present in the session and the individual's culture and personality. It is the task of the designers to construct the rules, roles, constraints, objectives and load in such a way that the essential elements for the aspect system of study are present.

2.6.2 Repeatability

The repeatability of gaming simulation is one of the pros of the method. In Figure 2.1 the inputs and outputs of a session are listed. The inputs on the left (rules, roles, objectives, constraints, load and situation) can be kept constant. For the situation this might be complicated when participants stem from different backgrounds or groups, but with some creativity the differences can be minimised, though should be documented. The participants will differ between sessions or will have the experience from a previous session which means that they are not the same participants as before. Gaming simulation is an excellent tool when a repeatable experiment is needed for different groups of participants / respondents and contextual aspects are important to know.

The type of experiment one can conduct with gaming simulation is close to a laboratory experiment. Klabbers (2008) gave three criteria to which analytical research with gaming simulation should adhere. They are: *control, randomisation and comparison*. These three criteria are standard in laboratory experiments. In gaming simulation, control is safeguarded through the manageability of the 6 groups of inputs (rules, roles, objectives, constraints, load and situation), via the load the selection of the participants, and via the situation the preparation of the participants. Randomisation can be safeguarded by randomly assigning roles to participants. Comparison can be safeguarded by keeping the 6 groups of inputs constant, which simplifies the comparison to a comparison between the outcomes, explained by differences in participants.

The relation between repeatability and validity is that real-world context is replaced by modelled context in a laboratory-style experiment. The validity of the real-world needs to be approached with the modelled context, thus losing some validity but gaining repeatability in the laboratory context. Important is that the human behaviour is not modelled but brought in with participants in the experiment.

2.7 Concluding remarks

Gaming simulation for analytical research is not new and hypothesis generation is the most frequently used function. Hypothesis testing is less common and this chapter placed this function in a methodological context. Literature (Klabbers, 2006, 2008) on gaming simulation in the design sciences opposes the use of analytical research methods. In this chapter it is argued that gaming simulation can be used as a laboratory-like environment to test hypotheses both qualitatively and quantitatively. The inputs are known and can be controlled. The uncontrolled variable is the group of participants, but their selection can be structured. Four aspects in the validity of experiments with this research method are distinguished: psychological reality, structural, process and predictive validity.

3. Research method in this study

The previous chapter positioned gaming simulation as a tool to generate and test hypotheses. The current chapter explains how this is done in the research in this book. The hypothesis generation and hypothesis testing functions each correspond with a cycle in the research process. The sensitivity analysis with a multi-agent simulation translates into two cycles.

The first section presents the research process, with subsections for each of the four cycles. The second section introduces the two gaming simulations developed for the two studies presented in this book. The chapter ends with concluding remarks in Section 3.3.

This chapter describes the finesse of the research process. For an understanding of the process in the following chapters it is recommended to read the first part of Section 3.1. The subsections describing the research process in each of the four cycles are meant as details for readers whose interests extend to a research process with gaming simulation, so they can, for instance, do it themselves or replicate the studies in this book.

3.1 Research process

The research presented in this book uses a research method that consists of four cycles: the design cycle, the empirical cycle, the multi-agent simulation cycle, and a support cycle: the multi-agent design cycle. Figure 3.1 shows an abstract representation with all phases connected, while Figure 3.2 shows the complex scheme of all detailed steps. In the detailed figure the four cycles can be recognised. As a rough introduction to the scheme one could say that in the

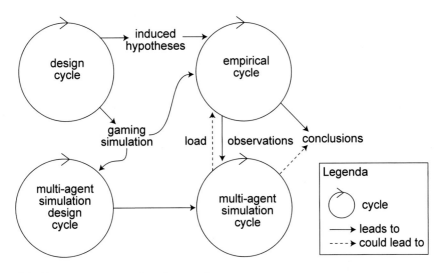

Figure 3.1. Abstract representation of research process.

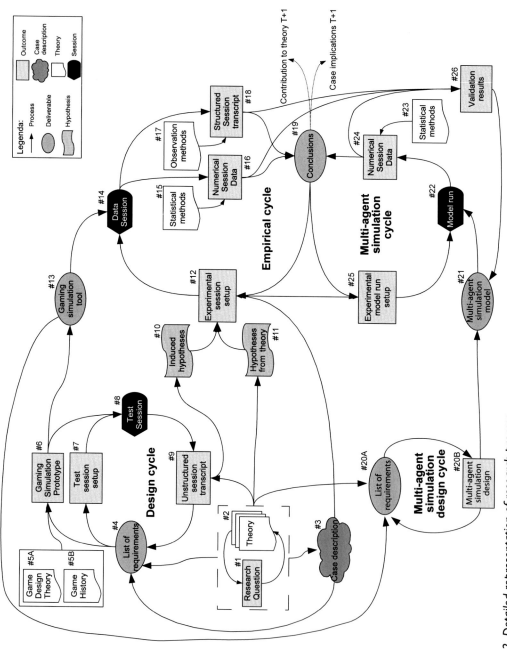

Figure 3.2. Detailed representation of research process.

The organisation of transactions

upper left design cycle the gaming simulation is developed and tested with first sessions. Then in the upper right empirical cycle the majority of the sessions are done with the tool developed first. Meanwhile in the lower left corner a multi-agent simulation can be designed with which in the lower right multi-agent simulation cycle runs can be conducted. Conclusions can be compared.

3.1.1 Integrated cycles

The design cycle results in two outcomes: firstly, there is the gaming simulation itself. It is input for the empirical cycle as the research tool and for the multi-agent simulation design as the reference system to model. Secondly, the design cycle results in induced hypotheses, which are input for the empirical cycle. The multi-agent simulation design cycle results in a multi-agent simulation model that is input for the multi-agent simulation cycle. The empirical and multi-agent simulation cycle both yield conclusions. Observations from the empirical cycle can be fed back into the multi-agent simulation cycle to research the conclusions in the other domain (human sessions in the empirical cycle or computer simulation in multi-agent simulation cycle). The multi-agent simulation cycle could lead to the selection of an interesting load (variable setting) to test in the empirical cycle. The last step has not been accomplished in this book, but is the reason that the proof-of-concept of the multi-agent design and simulation started. Conclusions can come from both cycles and should be checked against the other cycle.

3.1.2 Design cycle

As has been said in Chapter 2, using gaming simulation for research is like developing a new research instrument (the actual gaming simulation) for each specific research question. Of course, there are some gaming simulation frameworks (the so-called frame-games) that are able to answer a range of research questions, but even then the development of a playable gaming simulation takes time and tuning to tailor it to the specific research question. Therefore, a design cycle is needed to develop the actual gaming simulation to be played. An important by-product of the design cycle is a set of induced hypotheses.

The design cycle starts with the research question (#1). The research question leads to the selection of theories (#2) that can be used to answer the research question. In this book the main body of theory is new institutional economics, supported by theory on culture and bargaining (see Chapter 4).

The research question will be studied in the context of a case (#3). In this book there are two cases: a more generic network of goods with a hidden quality attribute and a real-world case in the Costa Rican mango supply network. The list of requirements (#4) is the list that determines what the gaming simulation to be designed should be able to achieve and what are the constraints. The research question determines what it should be able to achieve, while the case study provides the constraints of a particular real-world situation. The list of constraints contains the (theoretical or group-built) model to operationalise in the gaming simulation. Using theory on game design (#5A) and the rich history of gaming simulation that have been published about (#5B) the list of requirements leads to a first design prototype of the gaming simulation (#6). The approach for getting to a first prototype fast is similar to a design principle used in software engineering (Rapid Prototyping: Brooks, 1975). Using the prototype and a test session setup that determines a preliminary load and situation (#7), a test session (#8) can be conducted that results in an unstructured session transcript (#9). As one cannot predict the dynamics of a gaming simulation prototype in a session, the observer(s) aim to observe as much as possible, even though it might not seem relevant at first. Feedback from participants in a test session is included in the transcript. The unstructured session transcript can lead to induced hypotheses (#10).

There is structure in an unstructured session transcript, but this is not based upon theoretical foundations. When the design proceeds, the understanding of the unstructured session transcripts increases. In line with the grounded theory approach (Strauss and Corbin, 1998) induced hypotheses (#10) are constructed from the unstructured session transcripts. The induced hypotheses tackle the in-game behavioural hypotheses, while the hypotheses from theory (#11) handle the theoretical part.

Table 2.1 lists the relationship between the methodology in the current book, Grounded Theory and Duke and Geurts (2004).

Duke and Geurts (ibid) describe their methodology as a linear process. While it might be true that highly experienced game designers will reach the right solution quicker than beginner designers it is essential to test for the occurrence of the concepts and mechanisms that were supposed to be built into the gaming simulation in actual sessions. Experimenting with the load and the sensitivity of the outcomes for variations is essential and enhances the scope of data for building the induced hypotheses. That is why the methodology in this book uses a design cycle instead of a linear process. Other differences between the design cycle of Figure 3.3 and the methodology of Duke and Geurts are that phase II (Clarifying the problem) is summarised in all steps necessary to come to the list of requirements (#4). Furthermore their phases III and IV (design the policy exercise) is the process to get from the list of requirements to the gaming simulation prototype and the test session setup (#6 and #7). The game design theory (#5) contains Duke and Geurts' step 11 (techniques). The empirical cycle is very different from Duke and Geurts, as the gaming simulations fulfil a different role.

The unstructured game transcript is checked against the list of requirements to find out which of the initial goals and constraints does successfully appear in the prototype, what is missing and what is in the prototype that was not meant to be there. When the gaming simulation prototype aligns with the list of requirements, the transition can be made from prototype to gaming simulation tool (#13).

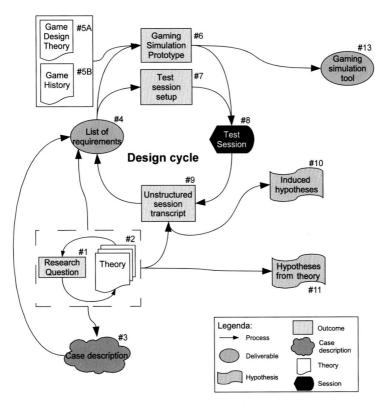

Figure 3.3. Design cycle.

3.1.3 Empirical cycle

The empirical cycle (Figure 3.4) can be entered when the gaming simulation prototype aligns with the list of requirements and when sufficient test cycles with the near-finished prototype have been conducted to be able to formulate induced hypotheses. It is not a necessity to use the induced hypotheses in the empirical cycle, as it is completely valid for testing theoretical hypotheses only. However, the process of grounded theory (description, conceptual ordering, theorising) is a prerequisite for a scientifically sound gaming simulation, as there needs to be an interpretation for the actions that participants can take within a session. A framing of the actions in relation to theory is required before entering the empirical cycle.

To come to an experimental session setup (#12), we need hypotheses. The hypotheses stem from two sources. Next to the induced hypotheses (#10) the theories used (#2) provide hypotheses based on existing literature (#11). The experimental session setup (#12) determines the load and situation. When needed, similarities or deviations can be used from the real-world situation in the case study (#3).

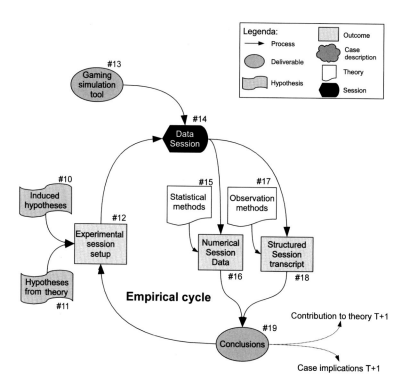

Figure 3.4. Empirical cycle.

Using the gaming simulation (#13), the experimental session setup leads to a data session (#14). The data session results in numerical session data (#16) and structured session transcripts (#18). We analyse the numerical session data with statistical methods (#15), and structure the session transcripts with observation methods (#17). The observation method used most is transcription of pre-defined events, like short-cutting the network, loud disagreements, traces, etc.

Using the quantitative numerical session data and the qualitative structured session transcripts conclusions can be drawn (#18) about the data session. The numerical session data describes what happened in the session, while the structured session transcript is used to give a real-world interpretation to the happenings. The conclusions provide information for updating the experimental session setup (#12) or for continuing with doing a similar data session. This depends on whether the numerical session data proves significant, or that the structured session transcript reveals new aspects that deserve experimentation.

When sufficient empirical cycles have been conducted to provide significant answers to the hypotheses, the conclusions are final and fed back as implications to the case study and contribution to theory. Now being at time T+1, a new project of design and empirical research can start on a new topic.

3.1.4 Multi-agent simulation design cycle

The gaming simulation (#13) is used in the empirical cycle to test hypotheses. The hypothesis formalisation phase cannot use a human gaming simulation tool, as the formalisation has the form of a computer model, more specifically a multi-agent model. When simulated, the multi-agent model should ideally show isomorphic behaviour to the human sessions in the empirical cycle. If so the multi-agent model is a validated formalisation of the hypotheses.

Two cycles are needed to come to such a validated model. First there is the multi-agent simulation design resulting in a multi agent simulation model. This model is based upon a list of requirements (#20A) that in itself is based upon the gaming simulation (#13) and theory.

Figure 3.5 gives a schematic representation of the multi-agent simulation design, resulting in a first version of a multi-agent simulation model. The actual process of coming to a multi-agent simulation design falls outside the scope of the current book. The work on this part has been performed in collaboration with multi-agent simulation modellers in a multi-disciplinary project.

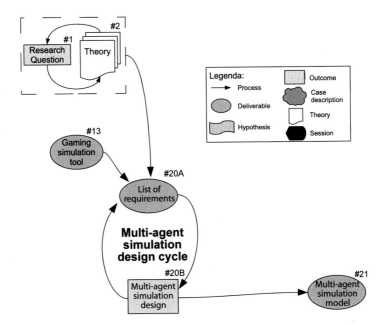

Figure 3.5. Multi-agent design.

3.1.5 Multi-agent simulation cycle

The second cycle needed to come to a validated multi-agent model is the actual multi-agent simulation cycle. Schematically this cycle is represented in Figure 3.6.

The multi-agent simulation model (#21) is run (#22). For the model run an experimental model run setup (#25) is needed that can initially be obtained from the human sessions (empirical cycle). A model run yields numerical session data (#24) that can be analyzed using statistical methods (#23). Session data can be compared with data from gaming sessions in order to validate the multi-agent model (#26). This can lead to adaptation of the task model or the decision functions or to the configuration of the model, or to the tuning of model parameter settings in order to better fit the gaming results. When the multi-agent model is validated and fits the outcomes of gaming simulation sessions, adaptations to the experimental model run setup can be made. In the combined research cycle, the mirroring of setups results in an experimental setup that makes comparable the variable settings for both game sessions and model runs. Conclusions from model runs can in combination with game session conclusions lead to refinements or falsification of theory, for instance improved or rejected models of decision functions.

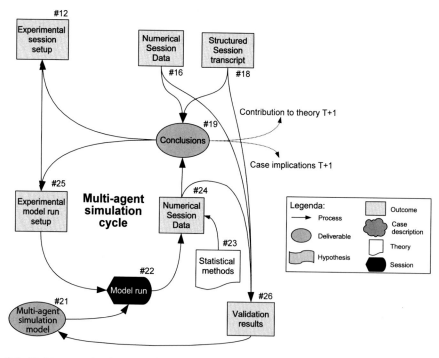

Figure 3.6. Multi-agent simulation cycle.

3.2 Two gaming simulations

As was mentioned in Chapter 1, this book presents two studies that together span the four sources of transaction costs (searching, bargaining, monitoring and enforcing). Each of the two studies had its own gaming simulation research tool (game in short). This section shortly introduces them, while more in-depth information can be found in Chapters 5 and 6, each of which presents one study.

3.2.1 Trust and Tracing Game

The first study uses the Trust and Tracing Game (TTG). This is a gaming simulation about the dilemma whether to trust or trace while trading goods with a hidden quality attribute. The TTG has been developed at Wageningen University with the initial purpose of using it as a learning tool. Using the TTG in this study was meant as a proof-of-principle as to whether it was possible to actually fulfil an empirical cycle and thus test hypotheses quantitatively. The TTG looked at the searching, monitoring and enforcing sources of transaction costs.

The research started when the TTG was in the last cycles of development. After previous test sessions the idea caught on to generate and test hypotheses with a gaming simulation, and the TTG was assumed to be a design that might fit this purpose. Section 5.1 describes the last iterations of the design cycle in which the gaming simulation was played with different situations and a small variation in the loads. Participants came from business and student populations. Insight was gained into the dynamics of sessions with the TTG. Section 5.1 ends with a list of variables to investigate further and some induced hypotheses. These have been used in Section 5.2 that describes the empirical cycle with the TTG. In addition to data from some of the last sessions in the design cycle, 27 additional sessions have been conducted. Section 5.2 tested two hypotheses, one from theory and one induced. Section 5.3 describes the multi-agent simulation cycle that began soon after the start of the empirical cycle and ended before the empirical cycle had been completed. Therefore, it is a proof-of-principle using observations of the design cycle and first empirical cycles.

3.2.2 Mango Chain Game

A second gaming simulation has been developed with the empirical cycle in mind from the start. This gaming simulation is called the Mango Chain Game. Its purpose was to test hypotheses in a development economics research trajectory of Zuniga-Arias (2007). Chapter 6 describes the design and the empirical cycle. No multi-agent simulation has been performed with this gaming simulation. The MCG looked at the searching and bargaining sources of transaction costs. Five sessions have been conducted with local smallholders in Costa Rica. The smallholders were mango producers for the export market.

3.3 Concluding remarks

This chapter described the way in which gaming simulation as a tool for hypothesis generation and testing is operationalised in two gaming simulation tools. The research process consists of four cycles, including two for sensitivity analysis using multi-agent simulation. Each of the two gaming simulations is used in a study in this book. The Trust and Tracing Game is further described in Chapter 5, and the Mango Chain Game in Chapter 6. The Trust and Tracing Game used all of the four cycles in the research process. The Mango Chain Game used the design and empirical cycle.

4. Reference theories

The main research question in this book (What is the influence of social structure on the organisation of transactions?) as was presented in Chapter 1 can be approached from many bodies of theory. Each of the two studies in this book contributes to this research question by analyzing empirical results of a particular dilemma in a specific supply network setting. Chapters 5 and 6 describe the specifics of the studies. However, the two studies share a theoretical framework and are both situated in the domain of supply chains and networks. The current chapter describes the shared theoretical background of both studies.

The organisation of transactions is the subject of study of multiple theories in the social sciences. Figure 4.1 shows a map of various theories from the social sciences grouped by their units of analysis (the human or the organisation). It should be noted that in the theories that analyze organisations the individual human is assessed as an agent representing a function, to be distinguished from the analysis of a human being.

In the domain of chains and networks the unit of analysis is the organisation, including the functions within the organisation. The theoretical foundations are found mostly in new institutional economics and network theory, with contributions from management studies and business economics. But to study the influence of social factors on the organisation of transactions in supply networks other theories are necessary that explain the human side of the people who fulfil a function in a chain or network.

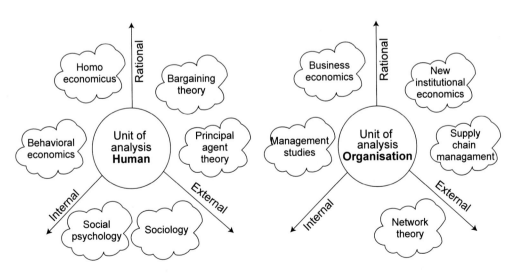

Figure 4.1. Explanatory theories grouped by unit of analysis.

The first section describes the theoretical background of the domain of supply chains and networks. Next, Section 4.2 goes into new institutional economics. The studies use a theoretical framework by Williamson (2000) to link theories. The last section (4.3) gives a map of where and how contributing theories are used in the studies in this book

4.1 Supply chains and networks

Supply chain management (SCM) is the integration of business processes from end user through original suppliers that provide products, services and information which add value to customers (Cooper et al., 1997). Lazzarini et al. (2001) describe two viewpoints used for analysing supply across the borders of a firm: the supply chain and the supply network perspective. The viewpoint of chains consists of a flow of goods from producer towards the consumer, a flow of money backwards and an information flow backward and forward (i.e. tracking and tracing). It pays strong attention to consumer orientation and chain responsiveness. The viewpoint of networks consists of relationships, alternative suppliers and buyers, and sectored collaboration.

Whereas early literature on supply chain management focused on the flow of goods (Cooper and Ellram, 1993), later papers indicate that SCM also includes: co-operative efforts between chain members in such areas as marketing research, promotion, sales and information gathering, research and development, product design and total system/value analysis. All functions or business processes need some level of upstream and/or downstream (vertical) coordination (Verduyn, 2004). The network perspective would add that some of these efforts need partners on the horizontal level too.

Lazzarini et al. (2001) try to integrate the two perspectives and introduce the term 'netchains' as an abbreviation for the often-used combination of 'chains and networks'. The term netchain has been used in the literature since then but it has not become widely adopted, though the concept of 'chains and networks' as two perspectives on the same phenomenon is common.

In the 2001 opening article of the first issue of the Journal on Chain and Network Science, Omta at al set a baseline for the discussion on contents and scope of chain and network theory. They cluster research in the domain into four main streams: network theory, social capital theory, supply chain management and business economics and organisational theory.

One of the main contributions to network theory is the actor, resource and activity model (Hakansson, 1992). The necessity for organisations to exchange resources is an important explanatory factor for inter-organisational relationships in this approach. Omta et al. (2001) state that in network theory, forms of collaboration are not just based on economic motivations; power and trust are key concepts in this approach (Uzzi, 1997). Kamann (1998) states that, based on the resource-dependence perspective, one can easily argue that neither buyers nor suppliers are completely free to select and change trade partners. The degree of dependency of an actor on its counterparts is contingent upon the criticality of the resources supplied (Chatterji, 1996). Power balances play a significant role (Omta *et al.,* 2001). Some of the trade relationships will be based on trust and loyalty, while others will be based on opportunism.

Social capital theory assumes that the behaviour and expectations of actors are constrained by the degree to which the relationship between the actors is embedded in the network structure. It is an important approach within the network theory (Omta *et al.,* 2001). Consequently, one may distinguish between a situation in which the network structure is closely knit and a situation in which non-redundant relationships prevail (Omta *et al.,* 2001). Coleman (1988) describes this phenomenon as the degree of closure of the network. Along the same lines, Granovetter (1985) introduces the notion of strong versus weak ties in a network to describe the depth of the relationship. Burt (1997) states that companies which are hardly if at all linked to other companies in the network form structural holes. Network relations may enhance the 'social capital' of a company, by making it feasible to get easier access to information and financial support. At the same time these relations may lead to 'social liability', e.g. by reducing the possibilities to relate to companies outside the network (Omta *et al.,* 2001).

Supply chain management aims at the integration of business planning and balancing supply and demand across the entire supply chain. It tries to bring suppliers and customers together in one concurrent business process. It spans the entire network from initial source to the ultimate consumer (Omta *et al.,* 2001; Tan, 2001). Planning and control are supported by modelling and ICT. Research in supply chain management is of a more technical nature than research in network theory. Control and coordination of stocks and material flows, exchange of information and the associated managerial and operational activities are the subject of study in this stream.

Business economics and organisational theory is rooted in the new institutional theory of transaction cost economics (Williamson, 1985) and agency theory (Eisenhardt, 1989). This stream is concerned with the rationale behind make-or-buy decisions. The theories explain the governance relationships of organisational cooperation, integrating views from business economics and organisational theory. In transaction cost economics the transactions between companies are the units of analysis. The three major characteristics of transactions are frequency, uncertainty and asset specificity.

This book is positioned in the business economics and organisational theory and the network theory stream. A framework from new institutional economics is used to link theories that explain the non-economic motives to collaborate from the network theory stream.

4.1.1 Sources of value

Dividing costs, benefits and risks in a network is not a zero-sum game (Hofstede *et al.,* 2004). When all partners collaborate for the common good, the network as a whole performs better.

Lazzarini *et al.* (2001, p. 8) identified six sources of value in chains and networks. Sources of value are 'strategic variables yielding economic rents. They can ... be associated with cost reduction, rent creation or rent capture.' They are:

Transformation. Assuming that a supply network is a single entity of management, the overall efficiency of production of goods and the operations needed to get these goods to the consumer (logistics, product-related information, demand forecasting, etc.) can be optimised. This should lead to a situation where more revenues are made for the supply network as a whole. (Simchi-Levi *et al.*, 2000)

Transactions. As networks do not have a hierarchy with central leadership, there are many transactions between independent firms before the product has reached the end-consumer. For each contract involving one or many transactions there are costs of searching, bargaining, monitoring and enforcing (Williamson, 1985; Coase, 1937). These transaction costs do not add value to a product and should be minimised through an appropriate governance structure.

Value capture. When one company in the network invests time and money to innovate, the revenues of this innovation are not necessarily earned back by this company if the innovation is easily copied by other companies (Teece, 1998). The whole network should work together to make sure it is beneficial for this company to invest, as the whole network can profit from it.

Social structure. This is a bucket category that could involve any aspect of relationships between people. As argued before, the more manageable sub-concept of embeddedness has been most studied by network researchers. Balancing embedded ties with arms-length relationships is important for the performance of the company (Uzzi, 1997). Having relationships with companies that the other people in the network don't know about has proven to be good for profits (Burt, 1992).

Network learning. This means developing knowledge jointly through collaboration. The term is from Diederen and Jonkers (2001). It can happen via two scenarios. First, there is the possibility of one much specialised company that teaches the other companies. This is known as knowledge spill-overs. (Kogut, 2000). The second possibility is the joint development of skills through which new practices emerge from combining individual skills and knowledge.

Network externalities. Network externalities occur when the benefits of adopting a technology, practice or contract increase with the number of adopters (Economides, 1996). Coordinating adoption in a network can create value for the network as a whole.

The research in this book contributes to the sources 'transaction' and 'social structure' by answering the research question that contains both sources. The *social structure* category is a broad bucket category with many contributing theories that explain interpersonal and business relations and the norms and values of the supply network as a whole. Trust between traders is the most prominent interpersonal variable. Business relations are expressed in a level of embeddedness as a measure of the density and the strength of ties between businesses. Norms and values in a network can be related to the culture traders come from. As *social structure* covers a broad range of theories, this book focuses on trust and embeddedness, with culture as an important context.

Dyer and Singh (1998) argue that 'an increasingly important unit of analysis for understanding competitive advantage is the relationship between firms...' They identify four sources of inter-organisational competitive advantage, i.e. relation-specific assets, knowledge-sharing routines, complementary resources/capabilities and effective governance. Effective governance can generate value by either lowering transaction costs, or providing incentives for value-creation initiatives, such as investing in relation-specific assets, sharing knowledge or combining complementary strategic resources (Dyer and Singh, 1998: 670).

Hofstede and Hofstede (2005) show that culture has its influences on almost any aspect of organised life that researchers have thought to investigate. Hofstede (2004) presents some implications for the coordination of networks. For instance, the attitude towards transparency, the tendency to switch trade partners and to select them from one's own in-group, the degree of acceptance of shared authority, the degree of willingness to engage in long-term obligations, and the degree of implicit trust can all be expected to differ with culture.

There is a multitude of aspects involved in judging the quality of the relations. Language and group identity, e.g. different studies, can divide people into groups. Culture (Hofstede and Hofstede, 2005), and more specifically the uncertainty avoidance factor, moderates the attitude of groups towards people from a different group. Do you trust people you don't know? And does it matter if somebody is from a different group?

Rousseau *et al.* (1998) show that economists, psychologists and sociologists tend to work with different conceptions of trust. This paper adopts the compromise definition presented by Rousseau *et al.* (1998: 395): *Trust is a psychological state comprising the intention to accept vulnerability based upon positive expectations of the intentions or behaviour of another.*

The keyword in this definition is *vulnerability*. Trusting people means that you do not need to take the trouble to check them, accepting the chance that they might cheat on you. Trust without vulnerability is gratuitous. This implies trust can only increase gradually through being tested in situations of reciprocal interdependency (Hofstede, 2003).

The importance of trust for supply networks is widely accepted (Harland, 1999). Camps *et al.* (2004) show that absence of trust is a reason for the failure of supply network projects. Trust is a key factor in being able to have a relationship (Claro *et al.,* 2004). Nooteboom *et al.* (1997) stress that trust enables partners to manage risk and opportunism in transactions.

Powell (1990) says that trust helps to reduce complexity in transaction making. Anderson and Narus (1990) explain that trust reflects the extent to which negotiations are fair and commitments are sustained. Uzzi (1997) shows that close relations (embedded relations as he calls it after Granovetter, 1985) with high trust are of key importance in the New York fashion industry. In New Institutional Economics trust becomes operational via the transaction costs. Uzzi (1997) showed that searching and monitoring complex transactions was less necessary with trusted partners.

4.2 New institutional economics

New institutional economics is a branch of economics that '... *provides a theoretical framework for understanding the trade-off that continuously occurs at the microeconomic level between alternative modes for organising transactions. (...) It also offers tools for analysing interactions between organisational forms of supply of goods (or "governance mechanisms" (...)) and the institutional environment in which they are embedded.'* (Ménard, 2000)

In this book the theoretical framework is based on Williamson (2000) and Menard and Shirley (2005). In his paper, Williamson discusses the state of his field (New Institutional Economics) by linking major contributions made by him or others. This book uses the 4-level framework presented in Williamson (2000) as the integrating theoretical framework. Menard and Shirley (2005) wrote 'The handbook of New Institutional Economics' in which he integrated views from thinkers like Williamson, Coase and North, among others.

Williamson (1998, 2000) offers a model (Figure 4.2) with four levels of analysis, tied to four time scales, in new institutional economics. The levels influence each other; higher ones set conditions for lower ones, while lower ones embody and change higher ones. Ménard's definition has a similar notion of different levels of analysis when he mentions the institutional environment that embeds governance mechanisms. Each of the levels changes about ten times as fast as the level above. Williamson calls the top level 'embeddedness', though this term is used differently from its use in other theory. As he lists customs, traditions, norms and informal institutions this level can be called 'culture'. Culture can be defined as 'the collective programming of the mind that distinguishes the members of one group or category of people from others' (Hofstede and Hofstede, 2005: 4). Culture in this sense is acquired in the early years of a person's life. Culture changes in centuries or even longer. Williamson (2000) calls the theory used on this level 'social theory'. Culture influences the second level, the institutional environment. At this level the formal rules of a trade community appear, often as the legislative environment of a country. Hence the theories used on this level are economics of property right and positive political theory. Changes occur in periods of tens to one hundred years.

Level 3 consists of the governance structure. Williamson (1996: 12) defines governance structures as ways of implementing order for facing potential conflicts that could threaten opportunities to realise mutual gains. Hendrikse (2003) is less abstract in his definition

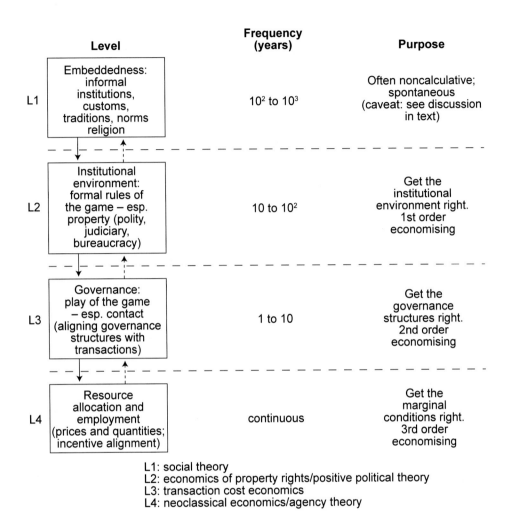

Level	Frequency (years)	Purpose
L1 Embeddedness: informal institutions, customs, traditions, norms religion	10^2 to 10^3	Often noncalculative; spontaneous (caveat: see discussion in text)
L2 Institutional environment: formal rules of the game – esp. property (polity, judiciary, bureaucracy)	10 to 10^2	Get the institutional environment right. 1st order economising
L3 Governance: play of the game – esp. contact (aligning governance structures with transactions)	1 to 10	Get the governance structures right. 2nd order economising
L4 Resource allocation and employment (prices and quantities; incentive alignment)	continuous	Get the marginal conditions right. 3rd order economising

L1: social theory
L2: economics of property rights/positive political theory
L3: transaction cost economics
L4: neoclassical economics/agency theory

Figure 4.2. Four-level model of Williamson (2000).

as a collection of rules, institutions and constraints structuring the transactions between the various stakeholders. Coase (1937) grouped arrangements of a structure under the expression 'institutional structure of production', while Williamson speaks of 'mechanisms of governance'. Menard (2005) captures the same ideas under the generic expression 'modes of organisation'. Menards term will be used here, as it makes clear the focus on the organisation of transactions.

There are three archetypical modes of organisation: market, hierarchy and network (Powell, 1990, Figure 4.3). The market mechanism is characterised by single-term transactions. Buyers and sellers constantly seek the best product for the best price and move to another trade partner if that is financially attractive. Costs and benefits are determined by the supply and demand curve and result in an optimal equilibrium price in case of a perfect market. The hierarchy mechanism uses contracts of some duration in which one party purchases production capacity from the other. Ultimately this means that the seller becomes an employee of the buyer. The network mechanism uses repeated transactions between independent companies or, more likely, between people in those companies. These business relations transcend the immediate business context, and shape mutual expectations on behaviour that will not harm the trade partner. A mode of organisation spans many transactions, whereas a transaction takes place under one mode of organisation.

Menard (2005) uses the name 'hybrid arrangements' as the third term instead of network, but mentions that this term is not fully satisfying. He sees hybrids as a range where '*at one end of the spectrum, close to market arrangements, hybrids rely primarily on trust: decisions are decentralised and coordination relies on mutual 'influence' and reciprocity. At the other end, hybrids come close to integration, with tight coordination through quasi-autonomous governing bodies or 'bureaus' sharing some attributes of a hierarchy.(...) Between these polar cases, mild forms of 'authority' develop, based on relational networks or on leadership. Relational networks (...) rely on tighter coordination than trust, with formal rules and conventions based on long-term relationships, on complementary competences, and/or on social 'connivance' (Powell et al., 1996).*' Both Williamson and Menard do mention the network (or hybrid) mechanism, but do not necessarily recognise it as a third type of organisation, like Powell. Their figures (and the name 'hybrid') suggest that it might be an intermediate between market and hierarchy that is not yet fully understood. Hakansson and Johansson (1993), Hakansson and Ford (2002), Burt (1982, 2000) and Coleman (1990) all recognise that a third mechanism exists. The current book uses the Powell framework of three distinct mechanisms because of its clarity, though does not reject the possibility that the network mechanism is a hybrid between market and hierarchy.

In the gaming simulations to follow in Chapters 5 and 6 only the appearance of trust-based and network-based hybrid arrangements is possible, to stay away from added complexity of trader-external institutions.

At level 3 the embeddedness that is discussed by Granovetter (1985) takes place. Granovetter distinguishes between theories that use under-socialised and over-socialised explanations of economic action within social theory. The concept of embeddedness, instead, acknowledges that ongoing networks of social relations between people discourage malfeasance. People guide their choices based on past interactions with people and continue to deal with those they trust.

At level 4 the actual business happens. Here is a constant flow of contracts and transactions going on. Transaction cost economics (Williamson, 1975) is a branch of economics concerned with the costs of doing business. The so-called transaction costs do not add any value to the product itself, but are the result of activities needed for doing business. These transaction costs should be minimised through an appropriate mode of organisation. On this level 4 the actual transaction costs are made, as (medium to short-term) contract negotiations and transactions cause the spending of time and money. The mode of organisation determines the structure of the transaction costs. And it is assumed that the mixture that provides the lowest transaction costs at level 4 of Williamson's analysis model will emerge at level 3. This emergence can be inhibited through restrictions from the institutional level (and indirectly the social embeddedness level) that embeds the governance. The histories of the organisations involved in the governance structure may cause path dependent outcomes, resulting in a different governance mechanism mix than the same situation without history would have resulted in (Ménard, 2004).

New Institutional Economics is principally concerned with levels 2 and 3 of Figure 4.2. But Williamson recognises that NIE cannot ignore level 1, although '*level 1 is taken as given by most institutional economists*'. At the daily level (4), the actions of actual human beings come into play. Social structure influences each of the four levels, with norms and values mostly at levels 1 and 2, and embeddedness and trust mostly at levels 3 and 4.

Figure 4.3 shows the modes of organisation. The corners are pure forms. The triangle spans a space of possible mixed forms. The axis between market and hierarchy can be characterised by the 'Make-or-Buy'-decision: do I buy on the market, or do I arrange production myself with the help of hired agents? Agency theory studies the axis between market and hierarchy. Within agency theory there are two streams of research: positivist and principal-agent models (Eisenhardt, 1989). The positivist stream focuses on identifying situations in which the principal and agent are likely to have conflicting goals and then describing the governance mechanisms that limit the agent's self-serving behaviour. This stream is most applicable to employer-employee relationships. The principal-agent stream is concerned with theory that can be applied to employer-employee, lawyer-client, buyer-supplier and other agency relationships (Eisenhardt, *ibid.*).

Williamson (1985) linked transaction costs and the mode of organisation through what he called the 'discrete alignment principle': traders will adopt the mode of organisation that fits better with the attributes of the transaction at stake. In doing so, Williamson provided a

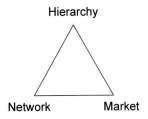

Figure 4.3. Modes of organisation (Diederen and Jonkers, 2001, after Powell, 1990).

way for empirical studies to circumvent the difficulty of measuring transaction costs directly, making organisational form the dependent variable (Menard, 2005).

For each contract between traders in a supply network involving one or many transactions there are costs of searching, bargaining, monitoring and enforcing (Williamson, 1985; Coase, 1937). These four factors are called sources of transaction costs. The sources of transaction costs provide a way to observe transaction costs in actual actions by observing the searching, bargaining, monitoring and enforcement actions of traders. Table 4.1 gives an overview of how the four sources of transaction costs can typically appear under three modes of organisation. This table is indicative and non-exclusive as a real-world mode of organisation is often a mixture of market, network and hierarchy.

Table 4.1. Typical appearance of transaction costs in three modes of organisation.

Searching	Bargaining	Monitoring	Enforcing
Market			
Investigation of prices for the product. Could be through auction or clearing house or from public information	In commodity market on price only. In other markets on product characteristics and price	By receiving information on quantity and quality. Regular check for adherence to transaction	Via legal institutions
Network			
First among companies with close ties, friends are preferred business partner	On all aspects of the transaction, plus possibly future transactions	Openness about process, information from other companies in the network.	Pressure from other closely tied companies. Possible damage to reputation. Focus on long-term relations in whole network
Hierarchy			
Non-existent as the leading company 'employs' the providing company to deliver.	Non-existent after settlement of initial contract between 'employing company' and 'employed company'	Of quality, quantity and process by the leading company	Leading company has power to determine payments or ultimately break the contract.

4.3 Other explanatory theories

In this chapter theories on supply networks and a theoretical framework from new institutional economics are discussed that are used in the two studies in this book. The following two chapters each present one study that uses the theoretical background and framework described in the previous sections. Both add specific explanatory theories that fit the specific hypotheses, dilemmas and context of the two studies. This section provides a map of how and where other theories are being used.

In Section 5.1 theory on culture and learning is added in the design cycle of the Trust and Tracing Game. In Section 5.3 the Homo economics from classical economics is used as a baseline scenario for behaviour. The endowment effect from Behavioural Economics is the subject of tests, as it is baseline behaviour to find. Chapter 6 adds bargaining theory.

5. The Trust and Tracing Game

The first of two gaming simulations used in this book is the Trust and Tracing Game (TTG). It is a paper-based game originally developed for learning about the relationship between social structure and the organisation of transactions in a trade network. This chapter describes the research performed using this gaming simulation. It builds upon three journal papers that were published in journals between 2006 and 2008. Each of the journal papers features one of the cycles of gaming simulation as set out in Chapter 2. Because of the structure of each of the individual papers the structure of the text does not fully match the numbered sequence of steps of the process as depicted in Chapter 3 as items are introduced in a manner suited for the paper.

The design cycle is presented in Section 5.1, in an adapted version of the paper by S.A. Meijer, G.J. Hofstede, G. Beers and S.W.F. Omta (2006), titled: Trust and Tracing Game: learning about transactions and embeddedness in the trade network. In: Journal of Production Planning and Control, Volume 17, number 6, 2006: 569-583. This paper describes experiences from 15 sessions with the gaming simulation. The gaming simulation model allows the use of network and market coordination mechanisms by participating groups. During debriefing participants typically indicated they learned that prior relationships were more important to the course of the session than economic theory predicts. Number of participants, language barriers, nationality, perceived group membership, and prior experience determined which transaction governance mechanism emerged in the gaming simulation. The type of experiment and independent variables in these sessions (as discussed in Sections 2.2 and 2.6.2) is that the game design is kept constant, and variations are made in the situation (business and student participants), some slight changes in load, and differences between participants (both in culture, level of education and background).

In Section 5.2, a reworked version of the paper 'The organisation of transactions; research with the Trust and Tracing Game' in Journal on Chain and Network Science, Volume 8, number 1, 2008: 1-20 by the same authors presented the empirical cycle. Specifically this paper presented the empirical results of research on the influence of social structure on the organisation of transactions in the domain of chains and networks. It shows results from 27 newly conducted sessions and previously unused data from 3 earlier sessions. Tests confirmed the use of network and market modes of organisation. Pre-existing social relations influenced the course of the action in the sessions. Being socially embedded was not beneficial for the score on the performance indicators for money and points. The hypothesised reduction in measurable transaction costs when there was high trust between the participants could not be found. Further analysis revealed that participants are able to suspect cheats in a session based on factors other than tracing. Testing hypotheses with data gathered in a gaming simulation proved feasible. The type of experiment and independent variables in these sessions (as discussed in Sections 2.2 and 2.6.2) is that the game design, load and situation has been kept constant (2 batches using explicitly differing loads, results merged because of no statistal differences found). The only variable input in these experiments were the participants.

Lastly Section 5.3 presents a proof-of-principle of multi-agent simulation, based on work originally presented in Dmytro Tykhonov, Catholijn Jonker, Sebastiaan Meijer and Tim Verwaart (2008): Agent-Based Simulation of the Trust and Tracing Game for Supply Chains and Networks, in Journal of Artificial Societies and Social Simulation vol. 11, no. 3 1. This paper describes a multi-agent simulation model of the Trust and Tracing Game. The multi-agent simulation can be applied to simulate the effect of models of individual decision making in partner selection, negotiation, deceit and trust on system behaviour. The combination of human gaming simulation and multi-agent simulation offers a basis for model refinement in a cycle of validation and experimentation. This paper describes a first round of model formulation and validation. The models presented are validated by a series of experiments performed by the implemented simulation system, of which the outcomes are compared on an aggregated level to the outcomes of games played by humans. The experiments systematically cover the important variations in parameter settings possible in the game and in the characteristics of the agents. The simulation results show the same tendencies of behaviour as the observed human games.

5.1 Design cycle

From its earliest design, The Trust and Tracing Game is a learning tool for investigating what drives the choice of governance mechanism in a trade network. It presents a simplified world of trade in a network. The gaming simulation confronts participants with the dilemma whether to rely on trust or to spend money on tracing when trading a product with a hidden quality attribute. The gaming simulation allows players to cheat a buyer to get a better price. Buyers can reveal cheats by requesting a trace.

The contribution of this section is to show that the Trust and Tracing Game provides an opportunity to make participants experience the influence of social structure on transactions. Through debriefing after a session the experiences are linked to theory. Participants learn about governance mechanisms. The Trust and Tracing Game has been played both in student and business sessions with groups of 9 to 20 participants. This section presents experiences from 15 sessions in the period 2003-2005 at the start of the project. In the methodological scheme this phase corresponds with the last cycle of the design phase, performing sensitivity analysis on a near-finished gaming simulation.

The section is organised as follows: firstly it argues that gaming simulations are useful for learning about transactions and embeddedness in supply networks. Section 5.1.2 introduces the Trust and Tracing Game. Section 5.1.3 presents results from the gaming simulation sessions done in this phase and links the results to theory from Chapter 4. The section ends with suggestions for further investigation, based on the preliminary work presented here. The suggestions for further research will be input for Section 5.2 as induced hypotheses.

5.1.1 Theory used in designing the gaming simulation

For creating the gaming simulation model, a number of bodies of concept were used as sources of inspiration. The article affords space to introduce each of them in brief. Their choice has been based on the author's perception of areas in which the state of the art about supply networks is still tentative. Inspiration was drawn from Ménard (2004, discussed at length in Hofstede, 2004). Ménard posits that the governance mechanisms of what he terms 'hybrid organisations' are still far from understood. He argues that trust, relational networks, leadership and formal government play a role.

5.1.1.1 Netchains: supply chains and networks

Supply networks are instances of Ménard's (2004) 'hybrid organisations'. The field of Chain and Network studies emerged in the late eighties (e.g. Powell, 1990) and has gained momentum during the last decade (e.g. Beers *et al.*, 1998; Omta *et al.*, 2001). It integrates two approaches towards studying supply networks, i.e. network analysis and supply chain analysis. The first approach is '...a broad field commonly associated with sociology but economists and strategy scholars have recently analyzed network-based industries and have applied network concepts to explain economic organisation and performance' (Lazzarini *et al.*, 2001). The second approach focuses '... on successive stages of value creation and capture in vertically organised set of firms' (*ibid.*). Lazzarini introduced the term 'netchain' to denote the combination of these two approaches.

5.1.1.2 Governance mechanisms

In any economic exchange, people have to agree on how to divide costs, benefits and risks (Williamson, 1985). The way in which the agreements are coordinated is called the governance mechanism. Three archetypal mechanisms exist, as displayed in Figure 4.3.

The market mechanism is characterised by single-term transactions. Buyers and sellers constantly seek the best product for the best price and move to another trade partner if that is financially attractive. Costs and benefits are determined by the supply and demand curve and result in an optimal equilibrium price in case of a perfect market.

The hierarchy mechanism uses contracts of some duration in which one party purchases production capacity from the other. Normally this means that the seller becomes an employee of the buyer.

The network mechanism uses long-term relationships between independent companies or, more likely, between people in those companies for control on (network) processes. These relations transcend the immediate business context. Relationships between people exist in myriad forms. Institutional economists tend to simplify the situation by making a binary distinction between instrumental or 'arm's length' ties, and close or 'embedded' ones (Granovetter, 1985; Uzzi, 1997). Uzzi (*ibid.*) asserts 'In the ideal-type atomistic market, exchange partners are linked by arm's length ties. Self-interest motivates actions, and actors regularly switch buyers and sellers (...)'. We shall adopt this definition and position arm's length ties in the market corner of Figure 4.3. When a group of people is linked by close mutual ties, that group is called an 'embedded network', from 'to embed', to surround closely. In an investigation of the clothes sector in New York, Uzzi (1997) found that ties were much more embedded than expected. Companies with embedded ties voluntarily exchanged information about, for instance, strategy, market information and operations.

Dividing costs, benefits and risks in a network is not a zero-sum game (Hofstede *et al.*, 2004). When all partners collaborate for the common good, the network as a whole performs better. Lazzarini *et al.* (2001: 8) identified six sources of value in chains and networks in a conceptual paper integrating supply chain and network analysis. Sources of value are 'strategic variables yielding economic rents. They can be either associated with cost reduction, rent creation or rent capture.' They are described in Section 4.2 of the previous chapter.

5.1.1.3 Social structure, culture and embeddedness

Because governance structures depend on relationships between people, it is reasonable to expect that they vary with the attributes of those people. To summarise this in the notion of embeddedness is to simplify dramatically. Aspects that could play a role include personalities, group dynamics, institutional environment, incentive structures, legal, economic and historical contingencies, and cultures. Taking all of these into consideration is not within the scope of this article. In this paragraph we shall highlight one aspect on which the sessions may shed some light: culture. By culture we mean 'the collective programming of the mind that distinguishes the members of one group or category of people from others' (Hofstede and Hofstede, 2005: 4). Culture in this sense is acquired in the early years of a person's life.

Williamson (1998) describes the dependency between governance structure and culture in a four-level model (Figure 4.2). Each of the levels varies due to external events, but each one does so at a time scale that is about ten times as slow as the previous one. Daily trade (level 4, days) happens in a governance structure (level 3, years) with agreements and long-term relationships between companies. The governance structure functions in the legislative environment (level 2, decades) of a country or region, or, with international trade, under international trade laws. The legislative environment is the operational result of the culture (level 1, centuries) of a country or region.

Hofstede and Hofstede (2005) show that culture has its influences on almost any aspect of organised life that researchers have thought to investigate. Hofstede (2004) presents some implications for the coordination of networks. For instance, the attitude towards transparency, the tendency to switch trade partners and to select them from one's own in-group, the degree of acceptance of shared authority, the degree of willingness to engage in long-term obligations, and the degree of implicit trust can all be expected to differ with culture.

It can be assumed that the cultural background of the participants is one of the factors that impact on their choice of coordination mechanism.

5.1.2 Learning using gaming simulations

There is a long history of teaching with gaming simulations (Duke and Geurts, 2004; Druckmann, 1994; Duke, 1974). In this subsection, the approach used here is positioned in the Kolb learning cycle, and the effectiveness of learning with gaming simulations is discussed.

5.1.2.1 Learning process

Kolb's learning cycle (Figure 5.1) is the standard conceptual framework when discussing learning processes. The learning cycle emphasises the sequence of experimentation, experience, reflection and conceptualisation. Gaming simulations follow the learning cycle once or several times, depending on the design of the gaming simulation. In case of a short,

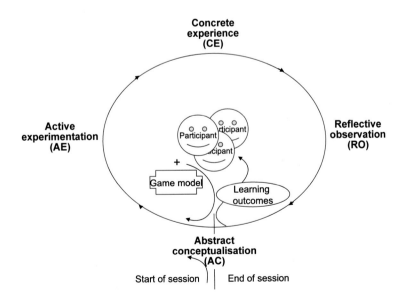

Figure 5.1. Kolb's learning cycle (after Kolb, 1984), with start and end of session, game model and participant.

one-cycle gaming simulation, the inner content of the learning cycle in Figure 5.1 describes the sequence of steps.

Before playing, participants are introduced to the gaming simulation design. They conceptualise what their task will be and how to win (AC). Then the session starts and participants experiment actively with the roles, rules and incentives of the gaming simulation (AE). The active experimentation leads to a concrete experience of the gaming simulation session (CE). During the debriefing participants reflect upon the experience (RO) and discuss how to conceptualise this (AC) in the group, moderated by a game leader. Learning outcomes are the spin-off of the debriefing.

5.1.2.2 Effectiveness of learning with gaming simulations

Gosen and Washbush (2004) studied the effectiveness of gaming simulations for learning. Summarising the best 39 papers out of 115 papers that describe results of gaming simulations, they conclude that at face value the effectiveness of both computer-supported simulations and experiential exercises (gaming simulations) is clearly supported by these papers. They do, however, criticise the lack of a shared measurement methodology to study the effectiveness. They even go so far as to conclude that: '*...the criterion variable being used, which is learning (...), is illusive. To our knowledge, every attempt to concretise this variable has failed. As a field, we believe we know what it is, but what it looks like so it can be measured lacks form. (...) The problem lies at least in part in the nature of learning and the learners' expressions of it. Learning is an internal mental process, and what is learned and how it is learned is unique for each individual. To create an instrument able to capture what is actually learned is easier said than done. To get at this, learners have to be motivated to express their learning,(...)*'. Most studies Gosen and Washbush found used self-reports of participants to indicate the learning. The second-largest group used objective measures. In this paper we use self-reports of the participants too. As the nature of the paper is explorative as regard what can be learned with the Trust and Tracing Game, objective measures cannot be determined in advance.

Druckmann (1994) investigated the effectiveness question too, and suggests that simulations with more 'fidelity' (similarity to a real-world situation) are better suited to learning actual practices, while more abstract simulations are better for learning insights. This is in line with the reasoning of Wenzler's diagram.

5.1.2.3 Gaming simulation about supply networks

Chapter 2 of this book positioned the approach used in this book to other gaming simulations and related approaches. At the time of writing of Meijer *et al.* (2006), the conclusion about the state of gaming simulation in the field of supply networks was that they are increasingly being used in the field, but learning results are still unknown. No results from gaming simulations about embeddedness and transactions were found in the literature.

The similarity between the nature of supply networks and the results from gaming simulations in policy making suggests that a gaming simulation approach could be eminently suited to learning about transactions and embeddedness in supply networks.

5.1.3 Trust and Tracing Game description

The Trust and Tracing Game is a gaming simulation. It models a generic food supply network. The product traded has a hidden quality attribute that can either be 'high' or 'low'. This could symbolise the invisibility of quality, freshness, harmful chemicals or pollutions, etc., appearing in food. There are four roles for participants: producer, middleman, retailer (these three roles are known as traders), and consumer. The game leader acts as the tracing agency for uncovering the hidden quality.

The theoretical fundaments on which the Trust and Tracing Game is built are discussed in Chapter 4 of this book.

Producers are supplied with products to be traded at the start of a session. They sell the products to middlemen. They are told which products are of high quality and which are of low quality, as this is not visible from the outside of the envelopes. Furthermore they receive an amount of change, a set of labels for 'high quality' and 'low quality', and a list of initial prices to think of when starting trade. The goal of producers is to make as much money as possible. The producer with the most money at the end of the session wins a prize.

Middlemen buy products from producers and sell them to retailers. They receive an amount of money, a set of labels for 'high quality' and 'low quality' and a list of initial prices to think of when starting trading. The initial prices are twice the prices of the producers to make sure transactions can start happening fast. The goal of middlemen is to make as much money as possible. The middleman with the most money at the end of the session wins a prize. Middlemen can add value to a product by tracing it at the tracing agency.

Retailers buy products from middlemen and sell them to consumers. They receive an amount of money, a set of labels indicating 'high quality' and 'low quality' and a list of initial prices to think of when starting trading. The initial prices are twice those of the middlemen to speed up the start of transactions. The goal of retailers is to make as much money as possible. The retailer with the most money at the end of the session wins a prize. Retailers can add value to a product by tracing it at the tracing agency.

Consumers buy products from retailers. Every product is worth an amount of points depending on the type and quality they receive. Consumers receive the list of points per product at the start of the session together with a budget to spend. Consumers must collect as many points as possible during the session. The consumer with the most points at the end of the session is the winner. Consumers are only allowed to check the hidden quality of their products between periods of the session. If they want to know the real quality during rounds they can request a trace from the tracing agency.

The **tracing agency** provides a means for anybody except producers to check the hidden quality of a product. Due to the labels that traders put on the envelopes when selling, he is able to see which trader sold the product for what stated quality. Thus he can reveal that there has been cheating about the quality. The tracing agency asks a fee from the trace requester if no cheat is found. If a cheat is found he fines the cheater and fines honest resellers of a cheated product to a lesser extent. Prices for tracing increase through the network: middlemen can trace cheaply as there is only one step to trace.

The model of the food product used in the sessions of the design cycle is a sealed envelope containing a note with the words 'high quality' or 'low quality'. Envelopes come in 3 types (visible on the outside), all available in high and low quality (not visible from the outside). Table 5.1 shows the number of consumer points for each envelope.

A transaction in the Trust and Tracing Game is an oral agreement between two participants about the trade of one or more products. Negotiated properties of a transaction are: the total price paid by the buyer and the amount, type and quality of the products delivered by the seller. Optional properties of a transaction are agreements on what to do when the seller cheats or has been cheated, conditions for future deals, and postponement of payment.

The risk per type of product changes with the difference in points for high and low quality products. Cheating for high-quality products of the blue type is more profitable than for the red type, which increases the temptation to cheat. The damage for the buyer increases too, for consumers are likely to pay prices in line with the number of points an envelope brings them.

Traders are never allowed to open the envelopes during a round. If they want to know the real quality they must ask the tracing agency for a trace.

Every trader must stick a label onto every envelope sold either from the high or low quality set of labels he has. It needs to be the label for the stated quality, i.e. the quality that he tells

Table 5.1. Points per envelope for consumers.

Quality\ type	Red	Yellow	Blue
Low	I	2	3
High	2	6	12
High:low ratio	2:1	3:1	4:1

his customer. The labels contain the name of the player (Producer1, Retailer3, etc) and a code that the tracing agency can use to know the quality. Traders in the network cannot see what has been said previously about the quality, but the tracing agency can see as he has the key to the code.

There are no other rules in the gaming simulation. Anybody is allowed to do business with anybody, although the physical setting (similar to the trade structure in Figure 5.2) suggests doing business with the adjacent traders in the network. This means that producers are sitting at the back of the room, each behind an isolated desk, the middlemen form a row with desks in front of them, and the retailers a row with desks in front of the middlemen. The consumers start sitting in front of the retailers. Anybody is allowed to walk around, but the first person in the neighbourhood at game start is the logical previous or next node in the network. Furthermore, during the explanation of the rules the only trade structure mentioned is the one from producer to middleman to retailer to consumer. The two explicit rules however do not limit people to undertake initiatives. (Golden rule #1 is: Do not open envelopes. #2 is: 'Do not do any real harm to each other.')

The session runs in three periods. Most sessions have periods of 20, 10 and 10 minutes respectively. Some sessions need more time when a group is slow in the start-up. There are two stop criteria. The first one is the moment when two producers run out of stock. This prevents price boosts due to supply shortage. The second criterion is when the normal time period is over and no new developments in the dynamics take place. The game leader observes negotiations and physical walks by participants to determine these dynamics.

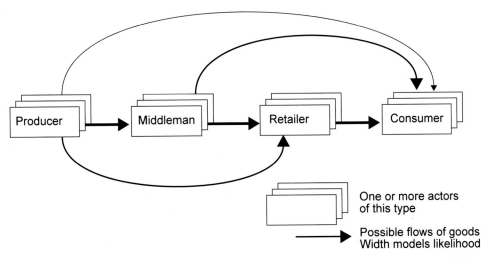

Figure 5.2. Possible flows of goods in the Trust and Tracing Game (Meijer and Hofstede, 2003b).

5.1.3.1 Appearance of theoretical concepts in the gaming simulation

What is being modelled?

The Trust and Tracing Game is in the abstract centre of Wenzler's classification. It models several important aspects of a trade network, but in a highly stylised way. The most important element modelled is like that in a real trade network, the actors are free to choose how to coordinate transactions. The gaming simulation is non-zero-sum thus the total revenues of all traders and the total amount of points for the consumers differ between gaming simulation sessions. The total revenues from trading grow when consumers pay high prices. For consumers who require a guaranteed (traced) product, the total costs will shrink when the actors in the network would work together to make the middleman trace the product. (Costs 2 euro, instead of 10 for the consumer). Total food safety goes up and transaction costs are lowest when neither cheating, nor tracing take place. Negotiation time goes down too when traders cooperate. Proper allocation of risks makes for smooth transactions and quick service to the consumers.

Notable omissions are that, contrary to real life, the gaming simulation uses one-person companies only, so that there are no company-internal dynamics. Furthermore, the gaming simulation has only one dramatically simple product (envelopes) and communication is all face-to-face in one room.

Network configuration and governance

The names of the roles and the physical setting of the participants in the session room suggest following the network structure of Figure 5.2 in trading. Reference prices in the players' instructions show an increase throughout the network. As described in the previous section, any interaction and transaction between participants is allowed. It is just neither mentioned nor stimulated. So the actual structure of the trade network is the emergent result of voluntary, partly unconscious choices by the participants. Elements of all of Powell's governance mechanisms could occur:

Market. Every session starts with an open market for products. There are no long-term agreements in advance and no hierarchies between companies.

Hierarchy. The gaming simulation does not allow for taking over another company and making the original company owner an employee. Pre-existing dominance relations between participants could, however, lead to *de facto* hierarchies.

Network. Embedded business relations can occur in the session. The design of the Trust and Tracing Game allows for any type of agreement, both implicit and explicit, between any numbers of participants.

Sources of value

Three of the six sources of value of Lazzarini *et al.* (2001) are operational in the gaming simulation. These are transformation, transaction and social structure.

Transformation. Producers are supplied with new products three times during the session, at the start of each period. Production itself is not part of the gaming simulation. One operation is possible: envelopes can be traced by the tracing agency. The quality information added to the product by this operation is considered to be part of the product, following Simchi-Levi *et al.* (2000). As traces are cheaper early in the network, consumers can optimise the costs of this transformation by having a retailer, or even better a middleman, request a trace. A retailer can gain by asking a middleman to trace. The trace requester can sell the traced product and have his tracing costs covered by the payment from his client who might appreciate the information added.

Transaction. Transaction costs in the gaming simulation occur in two mechanisms: time needed for negotiation and tracing costs. Long negotiations make for fewer deals during a session. Time pressure makes smooth negotiations economically favourable. Bulk trade instead of single products is another method for saving transaction costs per product.

The tracing costs only need to be paid by the trace requester when no problem with the product is found ('Cost of distrust', Meijer and Hofstede 2003b). When a cheat (or quality error) is found, the trader who caused it receives a publicly announced punishment. The public announcement possibly damages a trader's reputation. All resellers who did not discover the cheat and sold it honestly receive an 'ignorance' punishment. These are financial punishments discounted from the end-session balance.

Social structure. The embeddedness depends on antecedents that are not specified in the rules. The gaming simulation has neither an imposed hierarchy nor an imposed group structure. All players play for themselves and there are no multi-person roles. There are no formal restrictions on the interaction between players in different roles. Therefore participants never have to represent somebody else. If they do so, it must be because they stick to the trade sequence of the network.

If participants know each other beforehand, they take their relationships with them into the session. Real-world groups entering a session might not know each other personally, but could know the other's reputation from other participants. This history of existing relationships at the start of a session will influence the embeddedness depending on culture and the legislative and normative structure participants are used to. People who did not know each other or each other's reputation beforehand can establish new relationships.

Players can cooperate either by forming cartels, i.e. all producers working together (also known as horizontal cooperation) or by forming networks (a producer, middlemen and retailer working together, also known as vertical cooperation). With cartels access to markets can be blocked (by retailers and middlemen), or supply (producers) and demand (consumers) can be controlled to increase bargaining power. Networks facilitate guaranteed supply and demand by making participants stick to an agreement, which can facilitate trust from consumers in the quality delivered. Another use of networks is sharing of information about prices and costs and knowledge of supply and demand.

Culture and embeddedness

If the Trust and Tracing Game is played with multicultural groups, differences in attitude between national or professional cultures towards the value of embeddedness in trading can emerge. In line with Duke and Geurts (2004), the design allows the differences between participants to play a role in a session. This makes it feasible to learn about one's own attitude through confrontation with other attitudes.

Money and incentives

The currency in the TTG is the WURO; a fake currency hinting in the name to the euro, but worthless in the real world. The design was similar to money in the well-known Monopoly-game. Initial prices have been suggested in the instructions of participants, and tracing costs are fixed. The price level of goods however can differ between sessions, depending on the power distribution in the network and the amount of information participants have. Traders who do not know that consumers get 200 WURO may believe that their own start amount of money (15 WURO) is already a fortune. Trying to get as much information as possible could be a strategy to overcome.

The incentives between sessions can be compared even with the different price levels as care has been taken to make clear in advance that the producer who earned most money in the current session was the winner. The same announcement was made for middlemen and retailers. For consumers the person with the most points at the end of the game was determined the winner. It was made clear to the participants in advance that money left at the consumers after the session was worthless, and the same for envelopes (stock) at the traders.

5.1.4 Design cycle-sessions with the Trust and Tracing Game: method, results and analysis

Over the period 2003-2005 the authors conducted 15 sessions with the Trust and Tracing Game as a learning tool. The selection of the participants was not in our hands, as we were always invited to play the gaming simulation with pre-determined groups. Most of the sessions have been played with students in courses at universities and a primary school. The student session experiences are in the first subsection. Some sessions had participants from business environments or government. These experiences are described in Subsection 5.1.4.2.

A group gathered in a classroom or room in a meeting centre for a session of 3 to 4 hours, depending on the occurrence of an introductory speech from a lecturer. The rooms needed to be large enough to seat the participants behind a desk.

The typical flow of a gaming simulation session followed the cycle of Figure 5.1: the game leader introduces the dilemma and rules. Neither the topic to learn about nor the learning goal is mentioned upfront. Participants only bring in the (mental) models about trading they have from either theory or experience. Participants spontaneously discuss their strategy. Some acclaim loudly how they will deceive each other. Others try to form cartels with friends. Then the game leader assigns roles to the participants. This first goes on voluntary request, starting with the producers and ending with the consumers. Participants who did not ask for a role automatically get the role of consumer.

Next, participants move the tables in the room to form the trade sequence of Figure 5.2. Then the trading starts.

During two breaks participants get a chance to calm down and rethink their strategy. Consumers may check the quality of the envelopes they bought.

At the end of the session the game leader debriefs with the participants. Debriefings started with the question 'What surprised you in this session?' and ended with 'What did you learn from it?' The debriefing took the form of an open discussion between participants, moderated by the game leader. The results presented in the subsections below are the answers of participants to the two questions. The game leader helped groups that didn't spontaneously start discussing by mentioning conflicts or remarkable transactions observed in the session. The game leader noted the topics mentioned by participants for every debriefing. Comparisons are made between theory and the session experience and between real-life experience and the gaming simulation. Feedback from participants is welcomed. People can formulate their 'take-home' message and things learned.

Subsection 5.1.4.1 presents the results from the student sessions, while Subsection 5.1.4.2 presents the results from the business and government sessions. Subsection 5.1.4.3 analyses the results and links back to theories of Chapter 4. Table 5.2 provides an overview of the gaming simulation sessions.

Table 5.2. Overview of the gaming simulation sessions.

Code	Sessions	# Sessions
WUR	Student sessions Wageningen University, the Netherlands, course supply chain management	6
NSM	Student sessions Nijmegen University, the Netherlands, course supply chain management	2
KIDS	Primary school pupils De Boemerang, Apeldoorn, the Netherlands	1
MINAG	Ministry of agriculture, the Netherlands	1
PURDUE	Mid-career MBA students Purdue University, USA	3
TOMATO	Tomato growers and other network members, Germany and the Netherlands	2

5.1.4.1 Results in education

Few of the students in our sessions had actual business experience to refer to. Most obtained their knowledge about economics, markets and trade from books and courses. Six student sessions (WUR) were done at Wageningen University, the Netherlands (see Appendix A for details). Typical outcomes of these sessions were the following:

- In their courses students had learned about network forming, reduction of transaction costs and logistics optimisation. Participants used these terms during the introduction and start-up of the session.
- Students tended to forget about their strategy. They followed their feelings and often behaved opportunistically, especially the Dutch. During the two breaks most participants did not rethink their strategy. After the break they just went on with what they were doing: trading passionately.
- In the debriefing students were surprised that neither the rational decision logic of their course nor the economic model of a market functioned fully during the session.
- In international groups the differences between cultures in attitude towards uncertainty and cheating were eye-openers.
- Students expressed surprise at how their personal feelings about business partners influenced their purchasing decisions.
- The existing relationships between students did lead to preferred business partners in five out of six sessions. The non-Dutch students tended to seek one another out. The Dutch participants stuck to market governance mechanisms.
- In the session with only market mechanisms the game leader introduced the notion of embeddedness in the debriefing. Participants came up with situations in which they would have used their social relations to form embedded ties.

Two sessions (NSM) were conducted in the course 'Supply Chain Management' of Nijmegen University, the Netherlands. All participants were Dutch, and their study is management oriented. Sessions with these groups were in line with the Wageningen groups except for two conclusions in both sessions. Participants expressed surprise that low quality is not always worse than high quality. There is a market for low quality too; it is a different market segment than the one for high quality. A second difference was the surprise about producers having power since they have information, while the consumers have the money. Adding information to a product by tracing made participants reflect about 'transformation'. Most participants had been classmates for three years. Some of them were close friends. These close friends tended to favour each other as trade partners.

One session (KIDS) was conducted with a group consisting of fourteen children, 11 to 12 years old, in a primary school for pupils with learning difficulties. A few changes were made to the gaming simulation. Producers and consumers were restricted to stay on their seats, and the middlemen were omitted. The removal reduced complexity while retaining the dependency of retailers on their supplier for truthful delivery of products to the consumers. The group was able to play the Trust and Tracing Game instantly. Nobody in this group had any formal education in economics or trade. Half of the group came from families that were involved in trading in, for instance, used cars, second-hand computers or a shop at home. Children with this background were faster in trading, and could clearly explain afterwards why they were successful in the session or not. They recognised the use of auctioning, which has never been used by students, the use of alternatives by walking away to another supplier and the use of trust with friends. The other children were shocked about this. All of a sudden they realised how important friends in the group were. Existing social relations lead to embeddedness in the gaming simulation session.

5.1.4.2 Results in business environments

Early in 2003 a session (MINAG) was played at the Expertise Centre of the Dutch ministry of agriculture. Sixteen policy makers came together to learn about establishing successful networks in the agro-food business. Participants were between 45 and 65 years old and most of them had worked as policymakers for decades. As the participants were colleagues, they knew each other well. Some were friends.

A group of friends who happened to have sequential roles rapidly formed a network. Products from this network were trusted and never traced, contrary to products of the other suppliers. A producer with the notorious reputation of playing practical jokes on colleagues was particularly distrusted.

During the debriefing the policymakers reported their surprise at the extent to which personal friendships influenced their behaviour, rather than economic rationality. The distrusted producer turned out to be trustworthy after all, and the few consumers who had purchased his products had got good deals.

In 2003, 2004 and 2005 groups of managers in the American food industry visited the Netherlands for a two-week programme at Wageningen Business School as a part of an MBA programme of Purdue University. These three groups played the Trust and Tracing Game during a morning session (PURDUE) combined with a presentation on cultural differences in business. The participants had on average ten years of experience in a management position in a trade business. All had US nationality.

Each of the three sessions revealed a suspicious attitude of the participants towards their suppliers that resulted in massive tracing. In two sessions participants found out that setting up a network where the middlemen trace is the most economically efficient way to deal with consumers who want a guaranteed (traced) good.

In the debriefing the attention focused on two aspects. First, the differences between groups of different nationalities towards the uncertainty in the session were discussed. The PURDUE session solution was to make very explicit contracts, preferably written down and signed on paper and then to start trading with guaranteed products. Participants said they trusted the people with whom they had a contract.

The second aspect on which the participants drew conclusions was the benefit in the USA situation of having an early checking system that is trusted throughout the network. The optimisation of production of a guaranteed product this system allows was an eye-opener. Spontaneous discussion in the 2003 and 2005 session brought up 'network externalities' that subscription to the same trusted quality guarantee system could gain, although this is not present in the Trust and Tracing Game design.

In 2005 we played the Trust and Tracing Game with a large group of people from a tomato production network in the border area of the Netherlands and Germany. Half of the participants were Dutch, the other half German. Half of them were tomato growers, the rest consisted of seed producers, wholesalers and consultants. The group was split into two parallel sessions (TOMATO), each with an identical combination of growers and other real-world roles.

Both sessions ran very fast and participants bought large amounts of goods, up to the complete stock of producers at once. This had never happened in other sessions. The real-world wholesaler and a powerful representative of a research company were the ones who bought the large amounts. Producers mentioned in the debriefing that negotiation was impossible. The real-world hierarchy worked in the session too.

Except for one attempt by a Dutch grower, not one participant cheated. The buyers had a mutual complete confidence in their supplier. In the debriefing this was not a surprise to the participants until they realised that this is different in the national tomato market. This is a good trait of their network. Before the gaming simulation the network coordination project did not get the warm cooperation of the participants, but the gaming experience opened the eyes of the participants to the special quality of their group. Topics learned included insight into the low transaction costs within the group due to the high embeddedness as they know a lot about each other's companies.

5.1.4.3 Analysis of session experiences

What pattern did the sessions reveal as to the sources of value?

Transformation, by means of tracing, occurred never, with exception of the PURDUE sessions. This happened despite the fact that a lot of cheating occurred in most of the sessions. In the PURDUE sessions with their US buyers, sellers saw tracing as inevitable. They took the initiative and routinely used tracing as a marketing device. These were also the sessions in which the participants became quickly aware of the fact that tracing costs increased through the network.

Transactions turned out to be very dependent on social structure in all sessions. In almost all sessions, traders stuck to small quantities. Only in one session (TOMATO) did two of the most powerful real-world participants switch to bulk buying.

Social structure was most manifest in a number of variables:
- Number of participants. The level of energy and chaos increased with group size. This was no doubt partly due to greater feelings of anonymity from the game leader. But group dynamics have also very probably played a role.
- Language. Sessions tended to organise themselves along linguistic boundaries if their composition lent itself to such a division.
- Group identity. In the WUR sessions, all non-Dutch were of much more hierarchical, collectivistic and masculine cultures than the Dutch. And of course, only the Dutch were 'at home'. This contributed to a dichotomy of behaviour. The Dutch were more opportunistic and less careful of their reputation. Non-Dutch felt a common group identity and sought one another out. To what extent should these effects be attributed to culture, and to what extent to group identity, cannot be decided at present.
- Culture. The PURDUE sessions clearly differed from all the others in that tracing was used as a marketing device instead of a last resort in case of suspicion. This cannot be due to professional culture, or the TOMATO sessions would also have shown similar behaviour. It can be explained by the cultural difference between US (PURDUE sessions) and the Netherlands (dominant in all other sessions) (Hofstede and Hofstede, 2005). In an individualist, masculine culture such as that of the US, trust is never assumed and tracing is seen as normal behaviour. In a feminine culture such as that of the Netherlands, tracing is perceived as a sign of distrust, and avoided.

- Professional relationships. In the TOMATO sessions dramatically less cheating occurred than in all other sessions. The fact that this is an embedded network of people with real reputations at stake, who moreover were in a trust-building phase at the time of the sessions, can explain this. The session could damage real-world relationships, which means that the gaming simulation is not fully independent from the real world.
- Personal relationships. Many sessions included some participants who were friends and some who were strangers. The pre-existing friendships greatly shaped the trade network.

This means that social structure proved to be more than a one-dimensional factor. The word embeddedness does not do justice to the findings. It also implies that other variables were not obvious, e.g. gender and age. Duke and Geurts (2004) as well as Hofstede *et al.* (2002) emphasise the role of the game leaders in shaping a gaming session. All sessions described in this paper had the same game leader, except for one of the two TOMATO sessions, as these were run in parallel. The TOMATO sessions showed similar outcomes. This makes it impossible to draw conclusions about the role of the game leader based on the sessions described above.

How did the sessions function in terms of governance mechanism?

Hierarchy was apparent in the TOMATO session where real-world power wielders got the best of the market by imposing deals on the other participants. We observed that the real-world wholesaler offered the producers a decent and constant fee to deliver him all goods, regardless of quantity and quality. This is close to a job contract. In the other sessions there were no obvious real-world hierarchical differences. Many egalitarian networks occurred, but no single partner dominated this vertical integration.

Market mechanism was used by some of the participants in almost all sessions. These people spent some time keeping their ears and eyes wide open to get an idea of what others were charging. This market information was used to set prices. The price mechanism was the way to clear the market, as the lowest price won. Participants would not have perfect price information from trading with all possible partners, but the small physical distance from all traders, and the tendency of many traders to talk out loud with their potential clients made that participants who walked around could hear a lot of market information. However, a minority of traders actually traded with all possible partners, or shifted alliances during the session. Obviously this involved transaction costs that could be avoided by sticking to one partner. But pre-existing social structure was a more important predictor of the choice of trade partners than any other factor.

The **Network** mechanism was apparent in all but one session (WUR 3). Participants were quick to seize the most relevant cue of potential friendship or of group membership to pick their partners and they had a tendency to stick with those partners. The forming of embedded business relationships could be along the lines of friendships, or along cultural or linguistic divides.

Every session described above showed that participants recognised the influence of embeddedness on transactions, except for WUR session 3 where it was introduced by the game leader. In the debriefings the game leader introduced the governance mechanism triangle of Figure 4.3 (in all but the KIDS session) to structure the discussion. Comparisons between the session and daily life always led to the conclusion that good social relationships make for smooth transactions.

One related topic that was much debated was trust and its relationship with transparency. Transparency about the quality of the supply was discussed either as 'you should never inform your client that the quality probably is not right' or as 'you should always tell your client that quality isn't what he expects.' The PURDUE sessions showed a strong preference for institutionalised transparency with their guaranteed and traced products. This is in line with the dichotomised attitude to transparency described by Hofstede (2003): with friends, one can count on 'no news, good news' but with others, enforced transparency is desirable. This was apparent in the debriefing of the WUR, KIDS, MINAG and TOMATO sessions. In the US cultural environment of the PURDUE sessions, trust without transparency is not expected even by friends. The net result is that there was more cheating in the Dutch sessions. We cannot say anything about the influence of the transparency on profitability at this stage as first of all we collected qualitative information, and these sessions have been conducted in the last stages of the design cycle with (slightly) changing loads and situations. The profitability could also not be tested in mono-cultural US-sessions, as the value of money in a session is relative. Only tracing costs are fixed. We have seen differences in price levels between sessions, but within a session price levels always normalised to the same relative bandwidth. It would therefore require people with different transparency preferences in one session to make a comparison at this phase.

Uniqueness of sessions. As the session accounts (in Appendix A) show, no two sessions resulted in exactly the same learning. This can be explained by adopting the perspective of Duke and Geurts (2004). They consider the people who participate in a gaming simulation as an integral part of the dilemma that the gaming simulation is about. The people who play negotiate about what the problem is and what its boundaries are. Their different backgrounds, understanding of concepts and objectives mean that both the session itself and the debriefing afterwards depend on the people participating. This means that the exact learning outcomes cannot be fully determined in advance.

The managers of the PURDUE sessions were the best educated of all groups and they intensively worked on the sources of value during the MBA programme. This explains their skill at discovering economic mechanisms in the session. They even thought of a potential way to include network externalities by creating a joint quality guarantee system. Network externalities were not built into the gaming simulation as it stands.

In the student sessions the importance of real-life experience to the participants was shown, for example, in the jokes used after the game rules had been introduced. All student groups started to make jokes and remarks about what they plan to do using terms they recently learned in the course.

5.1.5 Data to collect in future sessions

Data collected so far does not provide answers to a number of specific questions about which variables drive the course of the session. Yet these questions would make the gaming simulation more useful and could increase potential learning. The results of the sessions give some pointers about which elements of social structure to take into account in collecting data about future sessions with the gaming simulation, in order to answer these questions.

Possibly relevant variables to measure *a priori* are: number of participants per session; number of participants per role, and for each participant: gender; nationality; age; profession; and for each dyad: degree of mutual acquaintance.

During the session, behavioural variables per participant X are relevant: with how many others has X traded; how trustworthy was X; if X requested a trace, why was that; what happened after a trace.

After a session, output variables can be collected. Per session: what was the price level? What was the speed? What was the quality level (as % false high-quality products)? Per participant: what financial result did X achieve, and what is X's reputation with trade partners?

5.1.6 Discussion and conclusions

The Trust and Tracing Game embodies an abstract, simple model of a supply network that can be played and debriefed within hours. In the gaming simulation all three governance mechanisms can be used, with emphasis on the network and market mechanisms. Participants in sessions with the Trust and Tracing Game have used the three mechanisms. This makes the gaming simulation valuable for learning about governance mechanisms.

The core dilemma of the gaming simulation is whether you rely on trust or spend money on tracing when faced with the possibility of being cheated upon. It was expected that real-world relationships influenced the embeddedness in sessions. The sessions presented in the paper clearly show this tendency. Participants learned about the value of embeddedness for smooth transactions in a supply network.

5.2 Empirical cycle

This section shows results from 27 newly conducted sessions and previously unused data from 3 earlier sessions. Tests confirmed the use of network and market modes of organisation. Pre-existing social relations influenced the course of the action in the sessions. Being socially embedded was not beneficial for the score on the performance indicators for money and points. The hypothesised reduction in measurable transaction costs when there was high trust between the participants could not be found. Further analysis revealed that participants are able to suspect cheats in a session based on factors other than tracing. Testing hypotheses with data gathered in a gaming simulation proved feasible. Experiences with the methodology used are discussed.

5.2.1 Rational baseline scenario

Gaming simulations in economics to demonstrate markets (like Holt, 1996) are most often accompanied by a calculation of the rational 'expected' outcome and the induced value of the commodity. As a 'what-if'-scenario the outcome of the TTG can be calculated that would occur if all participants strove for the rationally best outcome of the TTG. In such a scenario all envelopes would end up at the consumers, and all money at the producers, except for the start money and price for start stock of the middlemen and retailers. All middlemen and retailers would be out of business after selling their initial stock, assuming honest behaviour by producers.

The formula for calculating the induced value in WURO per point (in the load used in this section) is the following:

$$\text{Induced value} = \frac{\text{start money }(C) \times \#C}{(\text{start money }(P) \times \#P) + (\text{start money }(M) \times \#M) + (\text{start money }(R) \times \#R)}$$

Which under the current load works out as:

$$\text{Induced value} = \frac{200 \,\#C}{(156 \,\#P) + (4 \,\#M) + (24 \,\#R)}$$

Which under a regular distribution of roles (1:1:1:2) equals an induced value of 2.17 WURO per point. The end amount of money at each trader is:

$$\text{End money }(role) = (\text{start points }(role) \times \text{induced value}) + \text{start money }(role)$$

Where role can be producer, middlemen or retailer.

The end number of points at consumers is: E.

$$\text{End points (consumer)} = \frac{\textit{start money (consumer)}}{\textit{induced value}}$$

The complicating issue for a comparison between the session outcomes and this rational scenario is in the assumptions made:

- The good has a hidden quality attribute about which one can cheat. This makes the product not a real commodity, as not all characteristics are equal for any client. A client will value a product with a stated high quality differently depending on the trust he has in his supplier. A utilitarian approach to this valuation would be:

$$\text{Value} = (\textit{\%trust} \times \textit{HQpoints}) + ((1 - \textit{\%trust}) \times \textit{LQpoints})$$

- If the above had been the only complicating issue it would have been possible to calculate the value of trust by measuring differences between the prices paid and the rational baseline scenario, however there are more issues.
- A minor, though complicating issue is the impossibility to observe prices paid in all transactions. Only the end results can be calculated.
- The possibility to trace a product makes that there are two markets for high quality products: one for untraced products and one for traced products. Traced products will most likely be more expensive, as tracing costs are added. The tracing costs can be a measure for willingness to pay for (lack of) trust.
- As Holt (1996) mentions about the differences between his Classroom Pit Market game and the one developed by Chamberlin (1946) is that he observes a near-perfect market with equilibrium price because of the setting in which negotiators can hear each other. Chamberlain (1946) observed that the perfect competition model did not predict well, because he let people negotiate in distributed small groups. The TTG setting is more similar to the distributed Chamberlain setting where, as has been pointed out in Section 5.1.3, participants need to act to get information that is distributed in the room.
- Middlemen and retailers do not add strict economic value in situations where consumers do not want traced products. However, there are also mechanisms like loyalty to befriended traders, hesitance to bypass them and deviate from the suggested trade structure and service-oriented retailers who try to bundle consumer demand to be able to negotiate lower prices. As every session is unique, start-up effects and initiatives of individuals can shape the subsequent transactions.
- Therefore the TTG by design does not contain one market with associated assumptions that can potentially be a perfect market. Finding an equilibrium price with these many variables and settings is impossible. This implies that in the following tests of hypotheses the absolute price level cannot be a dependent variable.

5.2.2 From induced hypotheses to the empirical cycle

From observations of sample 1, Meijer *et al.* (2006) concluded that in TTG sessions the dominant mode of organisation is Network. Transactions were organised through repeated transactions with the same business partners, and sessions had been observed where trade was divided into language groups. To check this quantitatively in the new session, Hypothesis 5.1 is formulated:

Hypothesis 5.1: The dominant mode of organisation in the TTG is network, not market.

The design of the TTG means that every session starts by default with the market mode. Actual trading between people with possibly pre-existing relationships means that the network mode can emerge from subsequent transactions. The hierarchy mode is not accounted for in the design of the TTG, and can only manifest itself via pre-existing dominance relationships. The experimental session set-up avoided hierarchy through selection of the participants.

What indicators can be used to show whether the mode of organisation is network or market? Menard (2005) mentions trust and relational contracting as two drivers for the network mode.

Assumptions of a (perfect) market mode are:
- perfect information about supply and demand at no cost;
- a product that can be compared to any other item in the same market, i.e. a commodity;
- buyers will prefer the lowest priced item of two comparable products;
- no preference for a trade partner.

Contrasting these assumptions are the following assumptions of a trust and/or relational contracting-based mode:
- there are preferred trade partners for reasons other than price;
- trade depends on economic AND relational factors.

The assumptions lead to two tests that can confirm or reject Hypothesis 1:
a. there will be no preference for a trade partner;
b. the results at the end of a session depend on economic and relational factors.

Subsection 5.2.4 describes the tests and the outcomes.

Lazzarini *et al.* (2001) introduced six sources of value improvement for supply chains and networks. Meijer *et al.* (2006) state that the 'social structure' category was most manifest in the TTG in six variables: number of participants, language, group identity, culture, professional relationships and personal relationships. The social structure in a network consists of many relations between agents. There is a multitude of aspects involved in judging the quality of the relations. As said before, the professional and personal relationships are two viewpoints, but also language and group identity, e.g. different studies, can divide people into groups. Culture (Hofstede and Hofstede, 2005), and more specifically the uncertainty avoidance factor, moderates the attitude of groups towards people from a different group. Do you trust people you don't know? And does it matter if somebody is from a different group?

Rousseau *et al.* (1998) show that economists, psychologists and sociologists tend to work with different ideas of trust. This paper adopts the compromise definition presented by Rousseau *et al..* (1998: 395): *Trust is a psychological state comprising the intention to accept vulnerability based upon positive expectations of the intentions or behaviour of another.*

The keyword in this definition is *vulnerability*. Trusting people means that you do not need to go to the trouble of checking on them, accepting the chance that they might cheat on you. Trust without vulnerability is gratuitous. This implies trust can only increase gradually by being tested in situations of reciprocal interdependency (Hofstede, 2003).

The importance of trust for supply networks is widely accepted (Harland, 1999). Camps *et al.* (2004) show that absence of trust is a reason for failure for supply network projects. Trust is a key factor in being able to have a relationship (Claro *et al.*, 2004). Nooteboom *et al.* (1997) stress that trust enables partners to manage risk and opportunism in transactions. Powell (1990) says that trust helps to reduce complexity in transaction making. Anderson and Narus (1990) explain that trust reflects the extent to which negotiations are fair and commitments are sustained. Uzzi (1997) shows that close relations (embedded relations as he calls it after Granovetter, 1985) with high trust are of key importance in the New York fashion industry. In New Institutional Economics trust becomes operational via the transaction costs. Uzzi (1997) showed that searching and monitoring of complex transactions was less necessary with trusted partners. Being over-embedded however caused lock-in situations where it was impossible to do business or exchange information with new business partners. This leads to the formulation of Hypothesis 5.2:

> *Hypothesis 5.2: High trust between traders in a network reduces transaction costs.*

Both Menard and Shirley (2005) and Williamson (2000) believe that transaction costs are notoriously hard to measure. In the TTG one form of transaction cost (the checking costs) are modelled in a measurable way. The other forms of transaction costs are emergent and express themselves in the mode of organisation. To confirm the hypothesis while overcoming the measurement obstacle, three tests will be done.

a. test for correlation between (stated) trust and measurable transaction costs (Checking costs in the TTG);
b. test for correlation between (stated) trust and stated preferences for business partners;
c. test whether tracing was the only way to reveal cheats.

5.2.3 Experimental session setup

The load of a gaming simulation is the value of the initial configuration parameters a gaming simulation has. The situation of sessions are the characterising variables that are not part of the gaming simulation but can help interpret why sessions with similar people and a similar load may go differently. The load and situation of the sessions presented in this paper can be found in Table 5.3 and Table 5.4.

The data collection methods used consisted of a pre- and a post-questionnaire with questions using 5-point Likert scales. The questions are listed in Appendix B. Furthermore, the participants themselves counted game money and points at the end of a session and put their products (envelopes with labels marking transactions) and money in a participant-specific large envelope. The game leader made structured session transcripts during the sessions as a third source of data. His data is of a qualitative character and is used here in the discussion.

As the empirical cycle is of an iterative nature, better insights can lead to new ways of data collection or different loads or situations of sessions. In the case of the TTG the data collection of money and products hasn't changed. After a first iteration some questions were added to the pre- and post-questionnaire. The last series of sessions had a different load as a session consisted of only 1 round instead of 3. Appendix B lists the sessions conducted with the TTG. The *sample* column gives insight into changes between sessions. Sample 1 consists of all sessions used in

Table 5.3. Load of experimental sessions (P=producer, M=middleman, R=retailer, C=consumer).

Variable	Load A (sample 2 and 3)	Load B (sample 4)
Participants	8-24	
Division of roles	1P:1M:1R:2C. When perfect ratio is not possible: first add a consumer, then a producer, then a middleman.	
Rounds	3	1
Product	Sealed envelopes with different colour coding for type	
Start quantities of money (WURO)	P: 5 / M: 10 / R: 15 / C: 200	
Start quantities of products	P: 6 for every type and every quality (36 total)	P: same
	M: 2 low quality Yellow	M: none
	R: 2 high quality Blue	R: none
Tracing costs	M: 2 / R: 5 / C: 10, to be paid only when no cheat was found	
Cheat punishments	P: 2 / M: 5 / R: 10 plus public announcement	
Suggested prices	P: equal to the amount of points worth for consumers	
	M: 2.5 * prices for P	
	R: 6 * prices for P	

Table 5.4. Situation of experimental sessions.

Variable	
Selection of participants	Students in higher education, except for session 20. Participation was part of a course.
Real-world implications of participation	None. It was ensured by the game leaders that course teachers would not use results from a session in a grade.
Game leader	I, Sebastiaan Meijer for all sessions except 29 (sample I)
Duration of session	30 to 45 minutes of playing time. 2 hours max. including debriefing
Location	Classrooms of participating educational institute or similar venues in a conference centre

Meijer *et al.* (2006). Three of them could be used in other samples too as the experimental session setup coincidentally appeared to be the same. Sample 2 consists of all sessions with load A and a complete data collection. Sample 4 consists of all sessions with load B and a full data collection. Sample 3 consists of sessions that had load A but failed in data collection on one or more points due to external influences like no time for questionnaires, overly chaotic groups or inappropriateness of asking certain questions in a situation. The addition of extra questions in the questionnaires cannot be seen from the samples but expresses itself in item non-response when analysing sample 2. As this appears to play a very small role in the analyses in this paper sample 2 has been treated as one group.

Operationalisation of variables

Meijer *et al.* (2006) list variables to be taken into account when conducting research into factors that drive the course of action in sessions of the TTG. The results of their learning sessions (Sample 1 in Appendix A) gave some pointers about which elements of social structure to take into account in order to answer these questions. '*Possibly relevant variables to measure a priori are: number of participants per session; number of participants per role, and for each participant: gender; nationality; age; profession; and for each dyad: degree of mutual acquaintance. During the session, (the following) behavioural variables per participant X are relevant: with how many other has X traded; how trustworthy was X; if X requested a trace, why was that; what happened after a trace. After a session, output variables can be collected. Per session: what was the price level? What was the speed? What was the quality level (as % false High-quality products)? Per participant: what financial result did X achieve, and what is X's reputation with trade partners?*'

To be able to couple concepts of the theory and hypotheses to actual measurements this paper uses the analytical model in Figure 5.3 following the 4-level model of Williamson (2000) in

The organisation of transactions

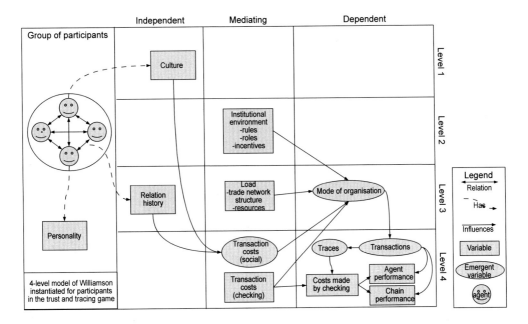

Figure 5.3. Analytical model for TTG, modelled after Figure 4.2.

Figure 4.2. Table 5.5 lists for each of the items in the analytical model how they are measured and using what tool.

5.2.4 Tests of hypotheses

From 27 newly conducted sessions and previously unused data from 3 older sessions a data set has been derived with 430 unique participants trading over 2500 product items in more than 5000 transactions. Appendix B lists the sessions. Sample 1-sessions are the older sessions used as empirical material for Meijer *et al.* (2006). Samples 2, 3 and 4 are new. Sample 2 and 3 share load A, while sample 4 used load B (Table 5.3).

All tests have been formulated before the data analysis started, however the tests in 2C were not planned for use with Hypothesis 5.2.

Hypothesis 5.1: The dominant mode of organisation used in the TTG is network, not market.

Test 1A: There will be no preference for a trade partner

In the Trust and Tracing Game every product can be expressed as a number of points for the consumers, and thus is a commodity. It is rational to assume that in the session where practical barriers between traders are absent, commodities will be bought from all available suppliers

Table 5.5. Operationalisation of variables for the Trust and Tracing Game (P=producer, M=middleman, R=retailer, C=consumer).

Variable	Measurement	Tool
Culture	Nationality as a proxy	Pre questionnaire
Relation history	Likert scale: average knowledge, difference in knowledge, and trust of other participants.	Pre questionnaire
Institutional environment	Rules and Roles enforced via game design. Incentives via orally announced award for best P, M, R and C.	Game design
Load	Structure: predetermined ratio between roles. Resources: amount of money given in advance, number of envelopes given in advance. Prices: via instructions	Game load: office preparation
Transaction costs (social)	Emergent, not measured. Expresses itself in mode of organisation.	Post questionnaire
Transaction costs (checking)	Price per check	Instructions
Mode of organisation	Share of buying and selling with each possible trade partner summed over all participants	Constructed from transactions
Transactions	For each product a list of sell-actions with the quality it has been sold for plus a mark for who ended up with the product.	Envelopes with labels.
Traces	For each envelope: has it been traced? If so: who traced?	Envelopes with manual trace marks
Costs made by checking	The number of traces multiplied by the trace fee per role of the tracer.	In the session: paid by tracer. Afterwards: calculated from traces.
Agent performance	For P / M / R: money left at the end. For C: number of points and amount of money left at the end.	Counted by participants and counted by game leaders afterwards. Game leader count was more accurate.
Network performance	Money per point, percentage of cheated envelopes. Percentage of product that reached the consumers. Distribution of profits over the network.	Constructed from transactions and performance agents.

equally. That is, when producers, middlemen and retailers will ask similar prices. For situations where for instance producers only consider middlemen or consumers as valid trade partners the situation is described below. For every actor in a session the 'selectiveness'-score has been calculated. The 'selectiveness'-score (SS) is defined as:

$$SS(Y_j) = \sum_{i=0}^{n} \text{sales}(X_i) \qquad \text{for i=1 to N}$$

Where:
Sales(X_i) = % of all sales with possible buyer X_i
X_i = each agent that could potentially buy from Y_m
Y_j = selling agent

Similarly the SS can be calculated for a buyer.

The theoretical SS for every actor can be calculated assuming equal trade with every trade partner possible. Assumptions have to be made as to who can be regarded as a possible trade partner. In the case of a producer there have been sessions in which the middlemen were the only trade partners, thus sticking to a strict network sequence, and there have been (many) sessions in which the consumers bought from producers directly. If the participants do not consider bypassing a node in the network to be appropriate behaviour, the number of possible trade partners falls. The rules of the TTG do not suggest or prohibit any bypassing. In the case of the TTG there should not be an automatic preference for consumers to buy from producers directly, only when they have the lowest prices. As middlemen and retailers in the TTG do not have any direct costs, like transport or production costs, they do not necessarily have higher prices. For traced products they even are the only source as producer cannot trace. In the qualitative sessions, described in Section 5.1. we observed that middlemen and retailers were trying to avoid consumer-access to the producers by making arrangements with the producers and being the first making deals with them. Middlemen tried to buy cheap and in bulk. Therefore there is no reason to assume, other than bad performance of middlemen and retailers that consumers only regards producers as valid trade partners.

In this analysis only the selectiveness scores for producers and consumers are calculated for two reasons. First the amount of consumers in the sessions was largest and from all traders the producers were the first role to get one person extra in case of asymmetric network configurations, therefore N is highest for these two roles. Second, being the start and the end of the network respectively, the SS is only one-sided. The structure of the data did not allow two-sided SS to be disentangled for supply and demand.

The theoretical minimum SS for producers assuming middlemen as possible trade partners is:

$$SS(P_j) = n \times \sum_{i=0}^{n} (1/n)^2$$

Where:
n = number of middlemen
P_j = producer

The theoretical minimum SS for consumers assuming retailers as possible trade partners is:

$$SS\,(C_j) = n \times \sum_{i=0}^{n} (1/n)^2$$

Where:
n = number of retailers
C_j = consumer

In Table 5.6 the outcome can be found for the non-parametric test for differences between the theoretical minimum score and the actual SS. The ranks table shows a majority of positive ranks for MiminumSelectiveness(M) – ActualSelectiveness, which means that the majority of the producers had an SS lower than the theoretical minimum. This indicates not only that the middlemen were a possible trade partner, but that all downstream agents can be considered. The MinimumSelectiveness(MRC) is the theoretical minimum SS for middlemen, retailers

Table 5.6. Wilcoxon Signed Ranks Test for selectiveness score (producers).

		N	Mean rank	Sum of ranks
MinimumSelectiveness(MRC) - ActualSelectiveness	Negative ranks	80	43.86	3509.00
	Positive ranks	4	15.25	61.00
	Ties	0		
	Total	84		
MinimumSelectiveness(M) - ActualSelectiveness	Negative ranks	24	30.96	743.00
	Positive ranks	57	45.23	2578.00
	Ties	3		
	Total	84		

Test statistics[c]	MinimumSelectiveness(MRC) - ActualSelectiveness	MinimumSelectiveness(M) - ActualSelectiveness
Z	-7.689[a]	-4.320[b]
Asymp. Sig. (2-tailed)	0.000	0.000

[a] Based on positive ranks.
[b] Based on negative ranks.
[c] Wilcoxon Signed Ranks Test.

and consumers available in the session of the particular producer. The ranks table shows only 4 positive ranks for MinimumSelectiveness(MRC) – ActualSelectiveness, and they stem from producers who did not trade at all. (ActualSelectiveness = 0). The test statistics show a 0.000 significance for ActualSelectiveness being smaller than the theoretical minimum. This rejects the proposition that there will be no preference for trade partners.

In Table 5.7 the outcome can be found for the non-parametric test for differences between the theoretical score (MinimumSelectiveness(R) and MinimumSelectiveness(PMR), respectively) and the ActualSelectiveness. The ranks table shows about 1/3 of positive ranks for MinimumSelectiveness(R) – ActualSelectiveness, revealing consumers who had an SS lower than the theoretical minimum. This, and the finding from the producers, indicates that consumers could buy from all upstream agents. The MinimumSelectiveness(PMR) is the theoretical minimum SS for producers, middlemen and retailers available in the session of the particular consumer. The ranks table shows 16 positive ranks that stem from consumers who did not buy at all. (ActualSelectiveness = 0) The test statistics show a 0.000 significance for ActualSelectiveness being smaller than the theoretical minimum. This rejects the proposition that there will be no preference for trade partners.

Table 5.7. Wilcoxon Signed Ranks Test for selectiveness score (consumers).

		N	Mean rank	Sum of ranks
MinimumSelectiveness(R) - ActualSelectiveness	Negative ranks	127	98.53	12,513.00
	Positive ranks	62	87.77	5442.00
	Ties	5		
	Total	194		
MinimumSelectiveness(PMR) - ActualSelectiveness	Negative ranks	178	104.60	18,619.00
	Positive ranks	16	18.50	296.00
	Ties	0		
	Total	194		

Test statistics[b]	MinimumSelectiveness(R) - ActualSelectiveness	MinimumSelectiveness(PMR) - ActualSelectiveness
Z	-4.696[a]	-11.701[a]
Asymp. Sig. (2-tailed)	0.000	0.000

[a] Based on positive ranks.

[b] Wilcoxon Signed Ranks Test.

Test 1B1: The number of points earned by consumers depends on economic and relational variables (assuming network mechanism).

A regression of the number of points earned by consumers with economic and relational variables yields the model in Table 5.8. (R-square = 0.486)

The relation between 'how many envelopes did you buy' and the number of points earned seems obvious at first, but considering the differences in points that each envelope is worth, it could have been the case that people buying only the 12-point-type had more points than people buying many of the lower-value envelopes. The number of points earned can further be explained by relational variables 'average knowledge of others' and 'trust in others', though negatively. The number of participants in a session is significant too. The number of suppliers negatively influenced the number of points earned. It appears from this analysis that the less you know the others and the more anonymous one is in a larger group, the more points you earn. Forming a good trade relationship with only a few suppliers indicates the use of the network mechanism. The amount of money spent is insignificant. This is in line with observations that prices differed enormously between transactions.

Table 5.8. Coefficients[a,b] of the number of points earned by consumers, sample 2, 3 and 4.

	Unstandardised coefficients		Standardised coefficients	t	Significance
	B	Std. error	Beta	B	Std. error
(Constant)	45	35		1.259	0.212
Money spent	0.11	0.087	0.12	1.280	0.204
# Participants in session	3.5	1.4	0.22	2.455	0.016
Average knowledge of others	-17	5.6	-0.26	-2.960	0.004
Difference in knowledge of others	-0.23	5.1	-0.004	-0.047	0.963
Trust in others	-12	6.2	-0.17	-1.974	0.052
Did you cooperate	-3.5	3.4	-0.08	-1.011	0.315
How many envelopes did you buy	4.8	0.71	0.79	6.808	0.000
How often have you been cheated upon	-2.0	1.7	-0.12	-1.217	0.227
How many suppliers	-7.2	3.8	-0.20	-1.899	0.061

[a] Dependent variable: points.
[b] IsConsumer = 1.00.

Test 1B2: The amount of money earned by traders depends on economic and relational variables (assuming network mechanism).

A regression of the amount of money earned by traders (Money) with economic and relational variables yields the model in Table 5.9. (R-square = 0.244)

For traders the number of envelopes sold is not significant for the amount of money earned. The number of buyers, the number of participants in a session and if they cooperated with somebody else determined the success of a trader. The insignificance of average knowledge and trust of other participants seems to contradict the model in Table 5.8, the differences of knowledge of others is significant. The cooperation factor upon closer inspection is influenced by producers cooperating and forming a kongsi. These kongsies were very successful in asking high prices. The significance of the number of people in a session suggests that the same more anonymous situation in which consumers earn more points works for traders too.

Table 5.9. Coefficients of the amount of money earned by traders, sample 2, 3 and 4.

	Coefficients[a]			t	Significance
	Unstandardised		Standardised		
	B	Std. error	Beta	B	Std. error
(Constant)	-165	55		-2.993	0.003
# Participants in session	6.0	2.4	0.20	2.470	0.015
Average knowledge of others	1.9	11	0.01	0.170	0.865
Difference in knowledge of others	14	7.8	0.13	1.724	0.087
Trust in others	11	11	0.08	1.001	0.319
Did you cooperate	13	7.2	0.14	1.770	0.079
How many envelopes did you sell	0.89	0.83	0.12	1.069	0.287
How many buyers	19	6.1	0.33	3.042	0.003
How often did you cheat	-3.8	2.4	-0.14	-1.557	0.122

[a] Dependent variable: money

Test 1C: Various correlations

Table 5.10 shows four correlations found in a correlation matrix of social variables 'Trust in others', 'Average knowledge of others' and 'Did you cooperate' with variables that described cheating, buying and selling behaviour. The matrix used data from sample 2, 3 and 4, separated in subgroups of 'traders' and 'consumers'.

Table 5.10. Various correlations for traders, sample 2, 3 and 4.

Variable		Corr	Sign.	N
Trust in others	% cheated of sell	0.170	0.029	165
Average knowledge of others	% low quality of total buy	0.150	0.046	178
Did you cooperate	How often did you cheat	-0.200	0.009	169
	How many envelopes did you sell	-0.221	0.004	169

Hypothesis 5.2: High trust between traders in a network reduces transaction costs.

Test 2A: Test for correlation between trust and measurable transaction costs (Checking costs in the TTG)

The outcome of this test is calculated using sample 2, 3 and 4. No correlation could be found between the stated trust level and the number of traces, as is shown in Table 5.11. When tested for each of the roles that were able to trace, there was still no correlation found.

Table 5.11. Correlation between stated trust and number of traces for different roles.

	M+R+C	Middlemen	Retailers	Consumers
Correlation	-0.085	-0.068	-0.167	-0.107
Significance	0.169	0.617	0.211	0.194

Test 2B: Test for correlation between trust and stated existence of a preferred business partner.

The outcome of this test is calculated using sample 2, 3 and 4. No correlation could be found between the stated trust level and the stated existence of a preferred business partner, as is shown in Table 5.12. When tested for each of the roles that were able to trace, there was still no correlation found.

Table 5.12. Correlation between stated trust and stated existence of a preferred business partner.

	M+R+C	Middlemen	Retailers	Consumers
Correlation	-0.062	-0.027	0.262	-0.201
Significance	0.505	0.891	0.206	0.105

Test 2C: Test whether tracing was the only way to reveal cheats.

It is invisible from the outside whether the product is of high or low quality. If the sub-propositions are confirmed, there have to be mechanisms at work between traders that make cheats detectable.

Test 2CA: The more a participant is cheated, the more he will trace.

Test for correlation between the number of traces and the number of cheated envelopes of a particular actor. Outcome: significant for retailers and consumers with sample 2, 3 and 4. Table 5.13 shows that the proposition cannot be generally accepted on a 5% confidence level. (Sign. = 0.085). The population consists of all cheatable participants, being middleman, retailers and consumers. When the population is divided into middlemen, retailers and consumers the proposition can be accepted for retailers and consumers.

Table 5.13. Correlation between number of traces and the number of cheated envelopes of a particular actor.

	M+R+C	Middlemen	Retailers	Consumers
Correlation	0.100	-0.085	0.263	0.187
Significance	0.085	0.510	0.034	0.015

Test 2CB: Tracers will reveal more cheats than random tracing would.

If cheating cannot be detected via methods other than checking an envelope, the chances of revealed a cheat with tracing are equal to the ratio cheated / not cheated envelopes the participant bought. The test shows significance for consumers using sample 2, 3 and 4 for all actors who traced for a difference in the means of the percentage cheated envelopes found in a trace and the percentage envelopes of all bought envelopes of each individual actor. The test is done both with a Wilcoxon Signed Ranks test to see whether the conclusion holds true for the assumption of similar distribution (Wilcoxon). This is significant for consumers on the 10% confidence interval (Table 5.14). For the other roles the number of tracers was too low to have a usable N.

Table 5.14. Wilcoxon Signed Ranks test for cheating detection (Ranks).

Role	PercentageCheatedOfTrace - PercentageCheatedOfBuy	N	Mean rank	Sum of ranks
Consumer	Negative ranks	19[a]	14.29	271.50
	Positive ranks	20[b]	25.43	508.50
	Ties	8[c]		
	Total	47		
Consumer	Z	-1.654[d]		
	Asymp. Sig. (2-tailed)	0.098		

[a] PercentageCheatedOfTrace < PercentageCheatedOfBuy.
[b] PercentageCheatedOfTrace > PercentageCheatedOfBuy.
[c] PercentageCheatedOfTrace = PercentageCheatedOfBuy.
[d] Based on negative ranks.

5.2.5 Discussion and conclusions

Test 1A confirms that there was a preference for some trade partners over others in the Trust and Tracing Game. This proves that the sessions were not a perfect market, because a perfect market would have forced the prices of all suppliers towards the equilibrium price and clients wouldn't have had preferences when the price was the same. Not reaching an equilibrium price is prohibited by non-perfect information for all participants, and the short setting during which a chain should emerge, but sometimes is bypassed. This bypass was most often not sudden, but occurred after some time of waiting of the consumers for retailers and middlemen to get them satisfactory goods. In several sessions there was a co-existence of consumers buying directly from producers and some buying from middlemen and retailers. This is most likely due to differences in personality that we did not test for. The design of the TTG makes that maing transactions is not about the product versus the price only. Buyers can require changes to the product (traced product) or may prefer to be served by middlemen or retailer to not bypass them.

Test 1B investigates what factors determine outcomes of the performance indicators money and points for the various roles. Included in the analysis are financial, relational and behavioural factors. Sub-tests show a difference between trader roles and consumer roles. Consumers gathered more points in sessions with more participants, when they knew and trusted the other participants less, when they had fewer suppliers and when buying with a bulk strategy. Neither the amount of money spent, nor being cheated upon, nor cooperating, nor knowing some people better than others were significant. The scores of the factors explaining the results of consumers led to the view that consumers who were successful used the network mechanism to set up a working trade relationship for repeated transactions in high volumes. Consumers who knew the others better and trusted them more were less successful in terms of points earned. Pre-existing relations thus hindered a fast exchange of goods in the Trust and Tracing Game setting.

For traders the situation was different. Successful traders were again in sessions with many participants, but they cooperated. The average knowledge and trust of others was not significant, but knowing some people better than others was. The more buyers they had, the better. Neither the number of cheats nor the number of envelopes sold significantly influenced the amount of money earned. From session transcripts and from the questionnaires it becomes clear that the preferred partner to cooperate with was somebody in the same role (especially producers among each other). The dominant mode of organisation used by successful traders seems to be market. Successful traders formed a monopoly on the products and traded with as many clients as possible. The importance of knowing some people better than others could indicate that the ones you knew better were easier to bind for a sell.

The outcomes of test 1B and the preferences for trade partners from test 1A led to the conclusion that pre-existing social relations did influence the course of the action in the Trust and Tracing Game. Consumers who earned many points were the ones that were less socially embedded, due to their lesser knowledge and trust of the others. They were able to form efficient network modes of organisation with a few suppliers. Traders who earned a lot of money used the market mechanism by forming monopolies and trading with as many clients as possible. Being socially embedded and letting your social network influence your trade behaviour was not positive for the score on the performance indicators in the Trust and Tracing Game situation.

Test 1C further illustrates this with correlations found. Traders who trusted others more cheated more on their buyers: a clear case of opportunistic behaviour. The better the average knowledge of the other participants the higher the percentage of low quality of their total buy, which is a way to avoid being cheated. Traders who cooperated more according to their questionnaire cheated less and sold fewer envelopes. Combined with the trader model from test 1B this indicates that cooperating traders were able to demand better prices for their goods, eliminating the temptation to cheat to earn more. Sticking to the market mechanism they just sold goods at a high price, without exploiting trust by being opportunistic.

The positive effect of the number of participants on the outcomes of both consumers and traders can be explained by the qualitative observations that in smaller groups the participants were watching each others' moves. People would listen to the negotiations of other people in the room. Fast exchange of goods was rare, especially in the beginning of the session. The larger the group, the more the noise levels went up, and hearing a conversation without being physically close was impossible. Participants who were a bit late in starting to trade quickly stood up and approached a possible trade partner. There were fewer passive participants in larger settings. A sound pressure level-measurement would have been an interesting variable in retrospect.

It has been impossible to find correlations between (ex ante) trust and the number of traces (Test 2A) or between trust and the existence of a preferred business partner (Test 2B). The trust stated in the pre-questionnaire did not show up in trade relationships. The next question is whether the tracing mechanism was the only source of finding out who was honest or not. Although technically it is impossible to know the real quality of an envelope until opening it, it might be that other mechanisms help in detecting cheats as well. Test 2C proves via two paths that a traced envelope was not a randomly selected product. There is a positive relation between how often one has been cheated and the number of traces (for consumers and retailers). This means that people who were cheated did suspect this and performed traces accordingly. A possible explanation is that the load A-sessions (Sample 2 and 3) were allowed to check their envelopes between rounds in the 3-round sessions. If they found out they had been cheated they could start tracing more. To investigate this explanation the second test in test 2C was carried out. It showed that the percentage of cheats found when doing a trace was higher than could be expected if a trace was a random choice. Because being cheated in the next round is independent of the round before, the second test proves that participants were better in detecting cheats than could be expected. There must be social mechanisms at work that detect if one is being cheated.

The initial assumption was that existing social relationships would be beneficial for building trade relationships, incorporating the social network in a newly formed trade network. In the setting of the Trust and Tracing Game the opposite is true when considering the performance indicators for money and points only. What has not been measured is the quality of the trade relations formed. A first possible explanation for the gap between the social relations and trade relations comes from the institutional environment the TTG provides. It might be that the people who were not successful in terms of the performance indicators for money and points were more successful in building a good relationship that might pay off in the future. The duration of a game session is relatively short. In almost any session the people who were less successful on the performance indicators explained afterwards that they were busy holding negotiations, exploring the wishes of their possible clients or waiting politely for busy producers to have some time to negotiate. Repeating the game with the same group of people would be very interesting in order to see if cheats from the past and the trade relations built would lead to new long-term relations with trusted partners. Changes in the incentive structure, for instance taking into account both money and the number of points for consumers and price per point earned by traders, would possibly change the nature of the negotiations. Sessions with a different incentive structure would be interesting for future research.

A second explanation for the gap between social and trade relations might come from the culture of the participants in the sessions. In the sample 1-sessions the multi-cultural sessions yielded interesting qualitative observations. In the sample 2, 3 and 4 sessions together 73% of the participants had Dutch nationality. On the cultural dimensions of Hofstede (Hofstede and Hofstede, 2005) the Dutch are particularly individualistic, extremely feminine, and have a lower than average uncertainty avoidance index. This combination of dimensions leads to a cultural profile in which the risk of being cheated upon is not very important to them, nor is it important to be a member of a permanent group. The dominance of the Dutch among the number of participants might explain the course of the sessions. The numbers of participants from other cultures are too low and they are spread over too many nationalities to distinguish quantitatively between nationalities in the analysis. A series of sessions with people from one or two nationalities could build a group to compare with the Dutch sessions. This way culture could be incorporated directly as an independent variable in the analysis.

A third possible explanation could be what Omta and Van Rossum (1999) called the 'dark side of cooperation'. In ten leading R&D firms they noted the negative effects of being embedded due to social liability, e.g. reducing the possibilities for relating to companies outside the network. Uzzi (1997) mentions the vulnerability of firms that are 'over-embedded' too, because they do not have access to new partners giving them a unique collaboration within the trade network. In the situation of the Trust and Tracing Game the incentive structure of the gaming simulation is strictly economic, as you either earn money or points. The social embeddedness could shape the transaction costs such that switching between trade partners is not likely. The possibility of a better price might not be attractive when you lose a friend and possibly a future business partner. Buying something might not be just for the purpose of points or money but could be a gesture to a friend too. While the Trust and Tracing Game allowed for making transaction costs a little more measurable via the checking costs, it cannot measure the other types of transaction costs in its current form. Remarks by Williamson (2000) and Menard (2005) to the effect that transaction costs are notoriously hard to measure still holds true.

Conclusions

The first hypothesis (*The dominant mode of organisation used in the TTG is network, not market*) could be confirmed for consumers and rejected for traders. Further analysis showed the influence of pre-existing social relations on the course of the action in the sessions. Being socially embedded was not positive for financial results in a session. Possible reasons for why this might be the case have been discussed. The second hypothesis (*High trust between traders in a network reduces transaction costs*) has been rejected. Neither measurable transaction costs from tracing nor the appearance of a preferred business partner with high trust could be found. Further analysis revealed that participants are able to suspect cheats based on factors other than tracing.

5.3 Multi-agent simulation

The reality of economic institutional forms is too complex to allow for direct individual participant-based analysis: there are too many people involved, the institutions influence each other, and nature plays an unpredictable role as well. An intermediate step between the real world and a model is needed that retains the essential elements of the economic institutional form under consideration. Special gaming simulations are developed that can fulfil the intermediate role (Duke and Geurts, 2004; Meijer and Hofstede, 2003a). By playing these gaming simulations with selected groups of human participants, the exactly same gaming simulation can be played in different settings, with people from different backgrounds, resulting in unique sessions. The so-called game load (variable settings) is kept constant, while the participants are varied through the so-called 'situation'. In this manner useful insight is gained into the way people behave in a certain dilemma (e.g. Zuniga *et al.*, 2007; Druckman, 1994; Van Liere *et al.*, 2004; Meijer *et al.*, 2006). However, the number of sessions that can be played with humans is limited as it is expensive and time-consuming to acquire participants (Duke and Geurts, 2004). Furthermore, one needs many sessions to control for variances between groups. In short, human gaming simulation is an essential step to overcome the complexity of real economic institutional forms. The number of sessions that needs to be played is a real disadvantage of the gaming simulation method.

The Trust and Tracing Game (Meijer and Hofstede, 2003b) played by human participants is used both as a tool for data gathering and as a tool to make participants give feedback on their daily experiences. Although the Trust and Tracing Game has been played numerous times (Meijer *et al.*, 2006), obviously, the problem of the number of sessions that can be and has to be played with humans and the expenses involved and the time-consuming nature of acquiring participants also holds true for this game. Therefore, of necessity, a research method had to be developed to overcome this problem.

The approach of modelling the Trust and Tracing Game differs from other approaches incorporating trust and supply networks in multi-agent systems. It models aspects of a structured socially rich trade environment in a fully automated agent model. TAC-SCM (http://www.sics.se/tac/) and the ART-testbed (http://www.art-testbed.net) are competitions of agents in a notional market that aim to find the best performing agent models, while the Trust and Tracing model is a research instrument to improve understanding of human behaviour in the game. The Global Supply Chain Game (Corsi *et al.*, 2006) uses models of a supply network in which multiple human agents can play. The emphasis is on network performance, where the Trust and Tracing Game focuses on human relations.

An empirically validated model of seller behaviour with regard to trust and deceit will be of value to the field of New Institutional Economics. A time-honoured method of checking whether a model is a correct representation of reality is by simulating the model and analyzing the results with respect to reality. However, as mentioned above, the problem of the real world is its complexity. Therefore, it is better to first study the phenomenon in the limited setting of a human gaming simulation, just focussing on the aspects that are well represented in the gaming simulation. The research method to understanding economic reality we introduce in this paper extends the idea of human game playing with agent-based simulation of those games and a rigorous validation method that incorporates both data from the human game playing and insights from conventional economic rationality. In our view agent-based simulation can to some extent overcome the disadvantages of gaming simulation in two ways. It can validate models of behaviour induced from game observations and it can be a tool in the selection of useful configurations for games with humans (test design). Validation of the models is done on the aggregated level using computer simulations. Simulation results are to be compared to a set of hypotheses based on human session observations (induced hypotheses) and conventional economic rationality (hypotheses from theory).

The theoretical contribution of this paper is in the introduction of an interdisciplinary research approach, bridging the gap between new institutional economics and agent-based economics by using human gaming simulation as an intermediary in which individual human transactions can be explicitly monitored and simulated. From the agent-technology perspective, the clear focus of a human gaming simulation presents detailed requirements for the design of the agent-based simulation, whereas the general aims of agent-based economics and new institutional economics set the general requirements of the agent-based simulation. The agent models introduced can be used as a point of departure for the modelling of other trade agents.

The practical importance of this work is that it provides a tool for research and education. Researchers can use the tool for research into trust and governance mechanisms in supply networks. Business schools and companies acting in supply networks can use the simulation and the game as training tools.

This paper presents the research method we developed, and illustrates its application for the study of supply networks, using the Trust and Tracing Game as gaming simulation. Section 2 describes the fundamental steps of the research method, The Trust and Tracing Game, and the results from human sessions. Sections 3 and 4 describe the foundations and elaboration of the agent model. The model has been tested for sensitivity to parameter changes with respect to trust level and honesty of the agents. Section 5 illustrates the validity of the approach by experimental results from multi-agent simulations. It presents the results of the sensitivity tests, together with a first validation against some hypotheses derived from the theory of new institutional economics and game session conclusions. The validation is on the macro level: tendencies expected from theory and game session conclusions correspond with tendencies in model run outcomes. Currently available data do not allow for model validation on the micro level. Section 6 presents the main conclusions of the paper and discusses future directions.

5.3.1 Research approach

The research subject is the role of trust and deceit in commercial transactions in different cultural and institutional settings. A main problem in the study of the mechanisms involved is that the macro-level system behaviour is not simply a linear combination of micro-level decision functions. There is great interdependence between the behaviour of actors. The main reason to apply gaming simulation as an instrument for experimental data collection is the opportunity it offers to collect data in controlled experiments on the effect of micro-level conditions on macro-level system performance.

Sessions played until 2005 provided many insights (Meijer and Hofstede, 2003b; Meijer *et al.*, 2006). We mention three examples applicable here:

- Dutch groups (with a highly uncertainty-tolerant culture; Hofstede and Hofstede, 2005) tend to forget about tracing and bypass the middlemen and retailers as they don't add value. This gives the producers a good chance to be opportunistic. The low tracing frequency encourages deceit.
- American groups tend to prefer guaranteed products. They quickly find out that the most economic way to do this is by purchasing a traced product and to let the middlemen do the trace, as this is the cheapest step. After initially tracing every lot, when relationships establish the middlemen agree with their customers to take samples.
- Participants who know and trust each other beforehand tend to start trading faster and trace less. The later indignation about deceits that had not been found out during the game is greater in these groups than it is when participants do not know each other.

Below we explain how gaming simulation and multi-agent simulation are combined to analyze the dynamics of the Trust and Tracing Game under different institutional and cultural settings. First, the gaming cycle is introduced. Then the combination of gaming and multi-agent simulation as applied in this research is explained.

In our approach we started with a ready-to-use gaming simulation that has been designed and tested in previous projects. It is ready for application in experiments with human subjects, to collect data for testing hypotheses. The data and conclusions from the experiments can be used to refine theory and formulate new hypotheses. The process of test design includes the variable settings for the experiments. Figure 3.4 presents empirical cycle.

The introduction of this paper referred to some shortcomings of the gaming simulation approach. The first reason for applying multi-agent simulation is that game sessions are time-consuming and require many new participants for each experiment. This research aims to provide a tool that in the long run can be used to select the most interesting game configurations to play. A second reason is the possible use of validated models to predict agent behaviour and test institutional settings and combinations of agents for their impact on supply network performance.

Figure 3.6 shows the cycle of a multi-agent simulation that complements the research cycle from Figure 3.4. By analysing the design of the gaming simulation, a task model for the agents is constructed. The decision functions implemented in these tasks are formulated on the basis of existing theory. Outcomes of model runs can be compared with gaming results in order to validate the MAS model. This can lead to adaptation of the task model or the decision functions, or to the configuration of the model, or to the tuning of model parameter settings in order to better fit the gaming results.

In the combined research cycle (), the process of test design results in an experimental setup that includes the variable settings for both game sessions and model runs. Conclusions from model runs can in combination with game session conclusions lead to refinements or falsification of theory for instance improved or rejected models of decision functions.

From the general theory of new institutional economics underlying the gaming simulation as explained in Meijer and Hofstede (2003b) and Meijer et al. (2006), and from the results of the first set of human gaming simulations, we constructed five hypotheses for a preliminary validation of the multi-agent model, following the cycle in Figure 3.6. The validation results are described in Subsection 5.3.4.

Some hypotheses refer to the *opportunistic, quality-minded*, and *thrifty* strategies defined in Subsection 5.3.4.2. *Opportunistic* traders aim to trade high quality for attractive prices, and are not particularly serious about the truthfulness of quality statements. *Quality-minded* traders do take these statements serious: like the opportunists they prefer high quality, but their priority is certainty, not attractive price. *Thrifty* traders prefer a good price and avoid risk, and have no particular preference for high quality.

- **Hypothesis 1:** When the initial willingness to trust is high the percentage of high-quality products sold is higher than when the initial willingness to trust is low.
- **Hypothesis 2:** In a homogeneous environment with all opportunistic agents there are more cheats than with other profiles.
- **Hypothesis 3:** In a homogeneous environment in which all agents are thrifty, i.e., who want to be certain about value for money, there are more traces than with other profiles.

- **Hypothesis 4:** Thrifty agents buy less high-quality products than opportunistic and quality-minded agents.
- **Hypothesis 5:** In a mixed setting with opportunistic and thrifty agents, the opportunistic agents cheat less than in a mixed setting with opportunistic and quality-minded agents.

5.3.2 Literature overview

The classical approach explains economic systems at the micro-economic (individual) level and at the macro-economic (system) level independently, using equilibrium-based models (McConnel and Brue, 2001). This approach is criticised for being unable to model various real-life economic and social systems such as financial markets and markets for fast-moving consumer goods (Moss and Edmonds, 2005). The new field of Artificial Economics (Batten, 2000) aims at building a bridge between micro- and macro levels through agent simulations that demonstrate how complex system properties emerge from the interaction of individuals.

Individual level models in the Trust and Tracing simulation model reproduce agents' decisions and behaviour in the following aspects:
- trust;
- deception;
- trade.

In the literature a variety of definitions of trust phenomena can be found. The common factor in these definitions is that trust is a complex issue relating belief in honesty, trustfulness, competence, reliability of the trusted system actors (e.g. Grandison and Sloman, 2000; Ramchurn *et al.*, 2004; Castelfranchi and Falcone, 2001; Jøsang and Presti, 2004). Furthermore, the definitions indicate that trust depends on the context in which interaction occurs or on the observer's point of view.

According to Ramchurn *et al.* (2004) trust can be conceptualised in two directions when designing agents and multi-agent systems:
- individual-level trust: agent's beliefs about honesty of his interaction partner(s);
- system-level trust: system regulation protocols and mechanisms that enforce agents to be trustworthy in interactions.

In this section we address problems and models for individual-level trust as our simulation environment already has system-level trust mechanisms, such as the tracing agency, that encourage trading agents to be trustworthy.

Defining trust as a probability allows it to be related to risk. Jøsang and Presti (2004) analyse the relationship between trust and risk and define reliability trust as 'trusting party's probability estimate of success of the transaction'. This allows economic aspects to be considered; agents may decide to trade with low-trust partners if loss in case of deceit is low.

With respect to deceit, our approach differs from that of Castelfranchi *et al.* (2001) and De Rosis *et al.* (2003) who treat deception strictly rationally as an instrument to win the game. In the social simulation aimed in our research we had to tune the agents to model actual human behaviour including their moral thresholds for deceit. Furthermore, our model does not simulate the purely rational decision as, for instance, the model of De Rosis *et al.* (2003) does.

Ward and Hexmoor (2003) describe an approach similar to ours, but their model does not explicitly recognise honesty as a moral threshold for deceit; it simply enables reinforcement learning from successful versus revealed deceit.

Our work acknowledges the work of Williamson (1998) stating that transaction cost economics possesses properties of bounded rationality, more precisely, that additional contractual complications can be attributed to an agent's opportunism rather than frailty of its motive. The importance of opportunistic behaviour is further supported by Diederen and Jonkers (2001) that mentions production quality and quality assurance as issues of chain and networks research to keep fast-switching consumers as a client.

In the real world networks, avoiding opportunistic (free rider) behaviour is an issue (Powell, 1996). Following the economic literature (Williamson, 1998; Diederen and Jonkers, 2001), the simulation has three economic incentives not to cheat:
- need to refund money (contract specific rules);
- fee from tracing agency (governance rules);
- damaged reputation / lowering of trust (social system rules).

The Trust and Tracing Game has possibilities to experiment with the relative importance and size of the three cost types. The contract-specific costs are easy to determine and therefore a calculated risk. The governance rules come with more uncertainty, because the fee depends on possible cheating of the agent you bought from. The damaged reputation depends on the socio-cultural system the agents come from, which formed their opinions on importance of trust and honesty and the reaction on being deceived.

5.3.3 Agent models

We apply a compositional design approach to the Trust and Tracing simulation. The components represent decision-making models for aspects of the agent's behaviour. The models of decision functions were partially published in Jonker *et al.* (2005a, 2005b). In this section we present the general architecture of the agent and explain details of the individual models. Additional information about the algorithms can be found at http://mmi.tudelft. nl/~dmytro/trustandtracing/.

5.3.3.1 Agent architecture

Types of agents acting in the simulated game are trade agents (producers, middlemen and retailers), consumers and the tracing agent. The architecture of the tracing agent is straightforward: it reports the real quality of a product lot to the requestor, informs the sellers that a trace has been requested and penalises untruthful sellers. In this paper we focus on the trading agents. The agent architecture for simulation of trading agents in the Trust and Tracing Game was originally described in Meijer and Verwaart (2005). For the research reported in this paper we apply the modified architecture represented in Figure 5.4: Trading agent architecture. All trading agents are built according to this architecture except the fact that producers do not have the *buying* process because they stand at the beginning of the supply network and receive products from the game leader and consumers do not have the *selling* process because they stand at the end of the supply network.

Trading agents start up with the initialisation process that handles communication with the game leader who informs them about initial stock and money. When the game leader broadcasts the 'start game' message to the agents, the initialisation process transfers the control to the goal and partner selection process. The goal and partner selection process decides to buy or to sell, depending on the agent role and stock position, and selects a partner at random, weighted by success or failure of previous negotiations with particular partners. Then the control is transferred to the trading process.

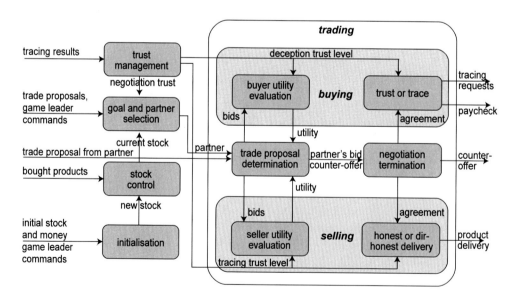

Figure 5.4. Trading agent architecture.

Because of its similarity to human bargaining behaviour, as evidenced in Bosse and Jonker (2005), we based the trading process on the algorithm presented in Jonker and Treur (2001). This approach to multi-issue simultaneous negotiations is based on utility theory. Negotiation partners send complete bids (a set of negotiation issues with assigned values) to each other. Once an agent has received a bid from the partner it can accept, or respond with a counter-offer, or cancel the negotiation that is decided in the negotiation termination process. Agents evaluate their own and their partner's bids using the buyer (seller) utility evaluation process that uses a generalised utility function that is a weighted linear combination of particular issue evaluation functions.

If agreement is reached the seller selects the product to be delivered to the buyer in the honest or dishonest delivery process (see 4.6). On the other side the buyer decides whether he wants to trace the product in the trust or trace process (see 4.7).

The utility functions involve individual experience-based trust as an argument for estimating risk. Modelling of trust for this purpose and experience-based updating of trust – as part of the trust management process – is the subject of the next subsection. In subsequent subsections we explain the utility functions and the way they can be used to represent agent's preferences or market strategies, and the decision-making models for tracing, delivery and goal determination.

5.3.3.2 Trust model

An important sub-process of the agent's trust management process is the update of trust values based on tracing results. Following Castelfranchi and Falcone (2001) we model trust as a joint subjective probability representing the opponent's willingness, capability, and opportunity to behave in a particular way. In the Trust and Tracing simulation we consider three different behaviours that we maintain a subjective probability about:
- successful negotiation (an agreement will be achieved);
- truthful product delivery (buyer: the opponent will deliver the agreed quality);
- not tracing the delivered products (seller: the opponent will trust after delivery).

Because all interaction that happens in the Trust and Tracing simulation involves two agents we model the trust as an individual-level agent's characteristic. This means that agent A can have low trust in agent B due to a series of bad experiences with him but have high trust in agent C, whose behaviour was honest and reliable.

In the Trust and Tracing simulation we assume that the experience gained by agents during the game is the only source of information about other agents. Thus, trust evaluation is built as a function of experience evaluation. Formula (1) formalises trust updating as a function of the agent's experience of trading with its opponent. Instead of Bayesian updating, in order to represent short memory and an endowment effect, we chose an asymmetric trust update function, with either completely positive or completely negative experience according to the classification of Jonker and Treur (1999).

$$\begin{cases} trust_{t+1} = (1-d)trust_t + d^+, & \text{if experience is positive} \\ trust_{t+1} = (1-d^-)trust_t, & \text{if experience is negative} \end{cases} \qquad (1)$$

where $trust_t$ represents trust after the t transactions. The value of $trust=1$ represents complete trust, $trust=0$ represents complete distrust, and $trust=0.5$ represents complete uncertainty. The model represents that the most recent experience has the strongest impact (short memory) and that negative experience may have a stronger impact than a positive experience. The latter is similar to the endowment effect (Hanemann, 1991). Losing trust that one thought to be endowed with has more impact than having a partner's trustworthiness confirmed. The factors d^+ and d^- are impact factors of positive and negative experiences respectively. They are related by an endowment coefficient e.

$$d^+ = e \cdot d^-, \qquad 0 < e \leq 1 \qquad (2)$$

5.3.3.3 Buyer model

Trade is an essential type of interaction between agents in the Trust and Tracing Game. In the Trust and Tracing Game trade is an agreement between buyer and seller achieved through negotiation. The negotiation issues are:
- type of the product;
- quality of the product;
- price;
- additional conditions: guarantee or certificate or none of these.

A buyer's motivation to accept or refuse a bid, depends on the price and other attributes of the bid, and on the player's trust in the seller. The buyer will compare the price with value of the product and decide. However, the value will depend on personal preferences of the buyer. Some buyers have a special preference for valuable high-quality products, motivated by some form of self-esteem; others prefer low quality to avoid the risk of being deceived. So the trade-off between value and price is not a rational decision in economic sense. A similar reasoning applies to uncertainty. Some players are prepared to evaluate the risk of being deceived based on their trust in sellers; others are afraid to be deceived and avoid risky transactions even if a rational economic evaluation suggests accepting the risk. Depending on the trust in the seller (belief about the opponent) and risk-attitude (personal trait of buyer), the buyer can try to reduce risk. Risk can be eliminated by trading low quality or demanding a quality certificate, or it can be reduced by a money-back guarantee. The attributes of a transaction are product type, stated quality, price, and certificate or money-back guarantee.

The negotiation model applied in this simulation requires that bids can be evaluated and compared in terms of utility. Real people will not actually calculate a utility. However, a utility function may approximate the player's preference in a simulation. For that purpose the utility is modelled as the sum of three terms, each having a weight factor. The three terms represent the price, the quality, and the risk associated with the bid. The weight factors may be used to model 'irrational' preferences for quality and certainty. The buyer's utility function is a weighted sum of normalised functions of price, satisfaction difference between high and low quality (for consumers) or expected turnover (for others), and risk (estimate based on trust in seller, guarantee and prices):

$$u_{\text{buyer}}(bid) = w_1 f_{\text{price}}(price_{\text{effective}}(bid)) + w_2 f_{\text{expected\_turnover}}(expected\_turnover(bid)) + \tag{3}$$

$$w_3 f_{\text{risk}}(risk_{\text{seller}}(bid))$$

The weight factors implement buyer's preference for a particular market strategy. For *quality-minded* buyers that are willing to pay to ensure high quality, both w_2 and w_3 are high relative to w_1, for instance $<0.2, 0.4, 0.4>$. The *opportunistic* buyer prefers a high quality for a low price but is prepared to accept uncertainty, for instance $<0.4, 0.4, 0.2>$. The *thrifty* buyer also prefers low price, but avoids risk, represented for instance by $<0.4, 0.2, 0.4>$.

The function of price f_{price} normalises the price of the bid according to the agent's beliefs about maximum and minimum market prices of the product of the given type and quality. The expected turnover normalisation function $f_{\text{expected\_turnover}}$ normalises the expected turnover with respect to the maximal and minimal possible number of satisfaction points. The risk of the buyer is normalised using the f_{risk} normalisation function which is based on the estimation of maximal risk over all possible bids. Such risk value corresponds to the bid with a 'yellow' product of high quality and the price equal to the agent's belief concerning the maximum price. This bid would lead to the maximal money loss in case of deception because the probability of deception attached to the seller does not change during the negotiation.

Effective price is the total cost of the purchase:

$$price_{\text{effective}}(bid) = price_{\text{purchase}} + cost_{\text{transaction}} \tag{4}$$

where $cost_{\text{transaction}}$ represents some extra cost for the buyer depending on the type of partner. In the current simulations the value is set to zero for purchases by consumers from retailers, by retailers from middlemen, and by middlemen from producers. It is set to infinity for all other combinations, to enforce the agents to follow their role in the supply network. In future simulations it may be varied to allow for the bypassing of some links. The expected turnover is the average of the agent's beliefs about the minimal and maximal future selling price of the commodity to be bought. For consumers the expected turnover is set to the satisfaction level.

The buyer's risk represents is calculated as the product of the probability of deceit and the cost in case of deceit.

$$risk_{buyer}(bid) = p_{deceit} \cdot cost_{deceit} \tag{5}$$

The probability of deceit is greater than zero only if the quality of the commodity quality is high and it is not certified. If these conditions are satisfied then the probability of deceit is estimated as the complement of buyer's trust in the seller.

$$p_{deceit}(bid) = q(bid) \cdot c(bid) \cdot (1 - trust(seller)) \tag{6}$$

where q=1 if the bid suggests high quality, 0 for low quality and c=0 if the bid suggests a certified transaction, 1 without certificate.

The costs in case of deceit are estimated for middlemen and retailers as the sum of the fine for untruthfully reselling a product and, only if no guarantee is provided, the loss of value that is assumed to be proportional to the loss of consumer satisfaction value. The formula for middlemen and retailers is:

$$cost_{deceit}(bid) = fine_{reselling} + loss_{reselling}(bid) \tag{7}$$

where

$$loss_{reselling}(bid) = g(bid) \cdot price_{effective} \cdot (1 - ratio_{low/high}(bid)) \tag{8}$$

and g represents the guarantee function (5): $g(bid)=1$ if the bid involves a guarantee; $g(bid)=0$ otherwise.

For consumers the cost in case of deceit is also assumed to be proportional with the loss of satisfaction value, but they do not risk a fine, so for consumers:

$$cost_{deceit}(bid) = g(bid) \cdot price_{effective} \cdot (1 - ratio_{low/high}(bid)) \tag{9}$$

This subsection presented the buyer's model. Before introducing the seller's model in Subsection 5.3.3.5, we present the model for the tracing decision entailed by a purchase.

5.3.3.4 Tracing decision

Tracing reveals the real quality of a commodity. The tracing agent executes the tracing and punishes cheaters as well as traders reselling bad commodities in good faith. The tracing agent only operates on request and requires some tracing fee. Agents may request a trace for two different reasons. First, they may want to assess the real quality of a commodity they bought. Second, they may provide the tracing result as a quality certificate when reselling the commodity. The decision to request a trace for the second reason originates from the negotiation process. This subsection focuses on the tracing decision for the first reason.

In human interaction the decision to trust or to trace depends on factors that cannot be modelled in a multi-agent system. Hearing a person speaking and making visual contact significantly influences the estimate of the partner's truthfulness (Burgoon *et al.*, 2003). So as not to completely disregard the variance introduced by these intractable factors, the trust-or-trace decision is modelled as a probability rather than as a deterministic process. The distribution involves experience-based trust in the seller and the buyer's confidence factor.

Several factors influence the tracing decision to be made after buying a commodity. First of all the tracing decision is based on the buyer's *trust* in the seller. Secondly, buyers may differ with respect to their *confidence*, an internal characteristic that determines the preference to trust rather than trace. It can be represented as a value on the interval [0,1]. We expect players with low trust to trace more frequently than players with high trust and we expect players with low confidence to trace more frequently than players with high confidence. Many other factors may influence the decision, like the amount of the tracing fee relative to the effective price, and the value ratio of low and high quality (satisfaction ratio). However, we have insufficient information to realistically model these unexplained influences. Therefore we modelled the decision to trust as a Bernoulli random variable with

$$p(\text{trust rather than trace}) = trust(seller(bid)) \cdot confidence \qquad (10)$$

If an agent has decided to trace the product, it sends a tracing request message to the tracing agent. Once the tracing result has been received the agent updates its trust belief about the seller and adds the product to the stock.

5.3.3.5 Seller model

The utility-based multi-attribute negotiation algorithm presented in Jonker and Treur (2001) is used to model the bargaining process. The seller's model mirrors the buyer's model. It accepts and produces bids with the same attributes:
- type of the product;
- quality of the product;
- price;
- additional conditions: guarantee or certificate or none of these.

To reduce the buyer's risk, a seller can give a 'money-back' guarantee if the product delivered turns out not to have the promised quality. A buyer can trace a product only when he has paid for it and received it. However, to completely eliminate a buyer's risk, a seller can request a 'trace' for the product which results in a certificate ensuring the real quality. The guarantee itself costs no money, but a certificate involves the tracing agency at the cost of a fee that depends on the position of the seller in the network. Tracing early on in the network is cheaper, as fewer steps have to be checked. Consumers pay the highest fee for tracing. Following Subsection 5.3.3, producers cannot trace, to force the environment to use at least one transaction with an unchecked product.

As explained above, we follow Williamson (1998) and Diederen and Jonkers (2001), in giving a seller three economic incentives not to cheat, which are included in the risk component of the sellers' utility function:
- Need to refund money. (Contract specific rules)
- Punishment by the tracing agency. (Governance rules)
- Damaged reputation / lowering of trust. (Social system rules)

The opportunity of deceit is not included in the utility function during negotiations. Firstly, for believable deceit sellers would have to act as if they were honest. Secondly, negotiation and delivery are separate processes and actual deceit takes place in the delivery phase. This resembles the real-world situation of firms having departments responsible for different functions in a firm. The model of the decision to deceive is discussed in the next subsection.

The seller's utility function is the weighted sum (linear combination) of normalised functions of effective price and seller's risk:

$$u_{seller}(bid) = w_1 f_{price}(price_{effective}) + w_2 f_{risk}(risk_{seller}) \tag{11}$$

The functions f_{price} and f_{risk} present the normalised effective price and seller's risk in the interval $[0; 1]$. The normalisation function of the price f_{price} is similar to the one of the seller. The risk is normalised over the maximum possible risk for the seller. This risk value corresponds to the bid of high-quality 'yellow' product with a money back guarantee and the price equal to the agent's belief of the maximum price for the product on the market. The weight factors add up to one and represent the seller's strategy with respect to the risk he is willing to take in reselling commodities of uncertain quality. In order to model a risk-neutral seller that acts rationally in the economic sense, $w_1=w_2=0.5$. For a risk-avoiding agent $w_1 < w_2$. We use $w_1=0.2$ and $w_2=0.8$ for the producers, middlemen and retailers following the quality-minded or thrifty strategies and $w_1=w_2=0.5$ for opportunistic agents.

The effective price represents the seller's benefit:

$$price_{effective}(bid) = price_{purchase}(bid) - price_{sell,min} - cost_{transaction} - cost_{certification} \tag{12}$$

where $price_{purchase}(bid)$ represents the proposed price of the anticipated transaction, $price_{sell,min}$ represents seller's belief about the minimal price he may receive from alternative buyers (opportunity cost); $cost_{transaction}$ represents the cost of making a transaction with the given partner, in the current simulations the value is set to zero for sales of retailers to consumer, middlemen to retailers, and of producers to middlemen (for more details, see the definition of the transaction costs in the 'Buyer Model' section); $cost_{certification}$ represents the fee a player has to pay the tracing agency for tracing the commodity and providing a certificate, if needed.

A seller's risk represents the risk losing money in case of reselling a high-quality commodity untruthfully delivered to the seller:

$$risk_{seller} = p_{negtrace} \cdot cost_{negtrace} \tag{13}$$

The probability of a negative trace is zero when the product is stated to be of low quality, or when the seller has bought the product with a certificate, or when seller would provide a certificate in the current transaction. Otherwise the seller has to estimate the probability that a trace would be requested (taking into account the trust he has in the buyer not to trace), and the probability that the product was untruthfully delivered by the supplier of the seller (based on the trust in the honesty of that supplier):

$$p_{neg.trace} = \left[1 - \prod_{seller_i \in S} trust_{honest}(seller_i)\right] \cdot \left[1 - trust_{tracing}(buyer)\right] \cdot q(bid) \cdot \left[1 - c(bid)\right] \tag{14}$$

where S is the set of agents upstream in the supply network of this particular lot; $trust_{honest}(seller_i)$ represents the experience-based trust the seller has in an upstream seller to deliver according to promise; $trust_{tracing}(buyer)$ represents the experience-based trust the seller has in its negotiation partner to accept a delivery without tracing it; $q(bid)=1$ if the quality is high, 0 if the quality is low; $c(bid)=1$ if a certificate is present or will be provided, 0 otherwise. Both $trust_{honest}(seller_i)$ and $trust_{tracing}(buyer)$ will be updated according to Equation 1.

'Money-back', governance fees and reputation damage are the components of cost in case of a negative trace. Whenever the player bought a high-quality product without a certificate and he again sells that product as a high-quality product without a guarantee, he runs the risk of a fee for untruthful selling, and a risk of reputation damage. If the seller provides a guarantee, there is the risk of having to pay money back and his reputation would be more severely damaged.

$$cost_{neg.trace}(bid) = fine_{good\ faith} + rep.damage +$$
$$g(bid) \cdot [rep.damage.guarantee + money\_back(bid)] \tag{15}$$

where $g(bid)=1$ if the bids entail a guarantee, 0 otherwise.

From the seller's point of view the 'money back' guarantee can be interpreted in terms of costs as an obligation to buy a low-quality product for a high-quality price: if seller is caught practising deception he has to pay the buyer the full price of the transaction but he receives a low-quality product back. In a formal way 'money back' can be considered as the following expression:

$$money\_back(bid) = price_{purchase}(bid) \cdot [1 - satisfaction\_ratio(bid)] \tag{16}$$

Reputation damage is difficult to estimate, because of the complexity of the phenomenon. It is currently represented in the simulation by a fixed amount of money, set by the game leader at a global level (reflecting societal values). However, further development of models is required. The Trust and Tracing Game simulation offers a good environment for developing and testing the models.

5.3.3.6 Honest or dishonest product delivery

The seller has to deliver a product after an agreement on transaction conditions has been reached. If low quality has been agreed, the seller will simply deliver a low-quality product. If high quality has been agreed, the seller may consider delivering low quality to gain profit. The decision to deceive is not merely a rational one with respect to financial advantages and risks. In real-world business social-cultural influences change the decision (Hofstede *et al.*, 2004). As mentioned before, we incorporate reputation and trust in our agents.

The opportunity of deceit occurs when the agent has sold a high-quality product without a certificate and has a low-quality product of the same type in stock. The motivation to deceive is in the extra profit that can be gained. In our model we assume that the motivation depends on the difference in consumer satisfaction between high and low quality. Three types of costs (money-back, fine and reputation / trust damage) provide a counterforce to the opportunistic behaviour, of which the third one comes from socio-cultural backgrounds.

In the agent model for delivery the trust level is only used to estimate the risk of being unmasked, so credulous buyers have an increased risk of being deceived. Thus, trust is not modelled as an incentive not to deceive friends, but only as an asset that enhances market position.

In reality other factors may influence the decision and not all of them can be taken into account. A random term represents the aggregated effect of unknown influences in the simulation. Furthermore, the random effect may cause some unexpected events that may prevent the simulation from reaching deadlock. The game leader can adjust the weight of the random term. Model calibration on human gaming data is necessary to find realistic values of this parameter.

All factors are normalised on [0, 1]. The following expresses the deceit decision.

IF $\quad q(bid) \cdot [1 - c(bid)] \cdot s[type(bid),low] \cdot$

$\qquad \{(1 - rtw) \cdot [1 - satisfaction\_ratio(bid)] \cdot trust_{tracing}(buyer) + rtw \cdot rnd\}$

$\qquad > honesty$ (17)

THEN $\quad$ deceive

where rtw is the weight of the random term, set in the interval $[0, 1]$; rnd represents a uniformly distributed random real number from interval $[0, 1]$; function s[type(bid),low] returns 1 if the selling agent has low-quality products in stock (opportunity to deceive), 0 otherwise. The temptation to deceive depends on the value ratio of low and high quality. Each agent has an honesty parameter that represents the agent's threshold for deceit.

Four parameters are used to model the dynamics of honesty. The first parameter is the initial level of honesty. The second parameter d^+ defines the honesty decay. Honesty is modelled to decay autonomously over time until some minimum level, which is the third parameter with respect to honesty. The fourth parameter is the tracing effect d^-. The awareness of being traced is assumed to improve honesty to an extent depending on d^-. The following equations model the honesty dynamics analogue to the trust dynamics in Equation 1.

$$\begin{cases} honesty_{t+1} = (1 - d^+)honesty_t + d^+ minimal\_honesty, & \text{if not traced} \\ honesty_{t+1} = (1 - d^-)honesty_t + d^-, & \text{if traced} \end{cases} \qquad (18)$$

5.3.3.7 Goal determination

In the game consumers will buy as much as they can. Producers will try to sell all products they have in stock. However, for middlemen and retailers some stock management is needed in order to negotiate efficiently. For example, imagine a situation in which a retailer tries to sell a product having in stock only a low-quality product with low satisfaction level. This would lead to a negotiation in which the consumer makes concessions in favour of the retailer, whereas the retailer proposes the same product in each bid. Such retailer's behaviour can break consumer's patience, leading him to cancel the negotiation and to update the negotiation trust accordingly. In such a situation it would not be reasonable for a retailer to start or enter negotiations as a seller.

Middlemen and retailers must decide to operate on the market as a seller or as a buyer. The decision to search a partner for selling or buying is taken at random with

$$p(sell) = \frac{\sum\limits_{i=\{Blue,Red,Yellow\}} \sum\limits_{j=\{Low,High\}} S_{ij}}{\sum\limits_{i=\{Blue,Red,Yellow\}} \sum\limits_{j=\{Low,High\}} T_{ij}} \qquad (19a)$$

$$p(buy) = 1 - p(sell) \qquad (19b)$$

where S_{ij} stands for actual stock level of product i and quality j, and T_{ij} stands for the corresponding target level.

In reaction to a proposal from a seller, a middleman or retailer will refuse if the stock of the proposed product is at target level, and enter negotiations otherwise. Expected turnover is set to zero in the buyer's utility of middlemen and retailers if the stock of the product/quality offered is at target level. Sellers will refuse negotiations if the requested product is completely out of stock.

5.3.4 Simulation results

This section presents the results of the first set of simulation runs, following the cycle in Figure 3.6. First, it presents the results of verification runs, aiming to test the model construction and the sensitivity for parameter settings (does it work as intended when designing the model?; do parameter changes adequately influence the results?). Subsection 5.3.4.1 presents the results of the simulation runs aiming at the preliminary validation of multi-agent model against the human gaming simulation based hypotheses that were postulated at the end of Subsection 5.3.2. All of the verifications and validations reported in this section concern aggregated game statistics; they do not concern the behaviour of individual agents. The results confirm that emergent behaviour observed in human simulation games can also be observed in the multi-agent simulations.

The agent parameters to represent individual characteristics of the agents are the trust parameters (initial trust and positive and negative trust update) and honesty parameters (initial and minimal honesty, honesty decay and punishment effect). Only these parameters and the trading strategies were varied in the results discussed in this section. Other parameters, such as game configuration, target stock levels, and fine and damage amounts, were equal for all simulations.

5.3.4.1 Verifying the multi-agent model

To test the multi-agent model we performed sensitivity analyses for the parameters that represent individual characteristics (trust and honesty). As a last verification we checked for the occurrence of the so-called endowment effect (Kahneman *et al.*, 1990). During the design of the multi-agent model the asymmetric speed of gaining trust and loosing trust has been a recurring issue. To test for correct implementation we assume that the endowment effect shall occur if trust is implemented correctly. Furthermore the results of this subsection give insight into the magnitude of effects when changing parameters.

All graphs in this section show lines that are interpolated between three data-points. Each of the data-points is based upon 3 runs, giving a 3*3 experimental setup, as shown in Appendix F.

Sensitivity for trust update parameters

The first series of simulation results demonstrate the effect of trust learning in an environment of perfectly honest traders. With an increasing value of the trust update parameters we expect the agents to learn more rapidly that they are dealing with perfectly honest traders. As a consequence we expect the proportion of high-quality transactions to increase, the proportion of certificates and guarantees to decrease, and the tracing frequency to decrease.

To test the actual sensitivity of the simulation model for trust learning, games were simulated with both positive and negative trust update parameters set to an equal value for all agents (three games with trust update set to 0.1, three with 0.5, and three with 0.9). The games were terminated after 500 transactions or as soon as one of the producers ran out of stock.

All agents were configured with a neutral negotiation strategy, assigning equal weights to transaction value and transaction risk, the latter being the product of estimated damage and estimated probability that the damage will occur. The weight tuples associated with this strategy are <0.33, 0.33, 0.33> for buyers and <0.5, 0.5> for sellers (see explanation of Equation 3 and 11). Each agent's initial trust in all other agents was set to 0.5, representing total uncertainty about the opponent's trustworthiness. All agents were configured to be perfectly honest (initial honesty and minimal honesty = 1.0). The confidence parameter was set to 0.95, so on average the agents will even request a trace in 5% of cases if they completely trust their opponents.

Figure 5.5 presents the statistics of the simulated games (see Table A1 in Appendix F for details). The effects of increasing trust update factors on statistics in games with perfectly honest agents are:
- increasing proportion of high-quality transactions (H/N=0.64, 0.74, 0.78);
- decreasing proportions of certified or guaranteed transactions (C/H=0.10, 0.06, 0.05; G/H=0.77, 0.54, 0.50);
- decreasing tracing ratio ($P/(H-C)$=0.40, 0.20, 0.13), so game statistics are sensitive to the trust update factor as expected.

Sensitivity for initial trust

Table A2 in Appendix F shows statistics of simulated games with different strategies that are defined in the explanation of Equations 3 and 11, different values of initial trust, and different values of honesty. All games were simulated with a homogeneous agent population: in a particular game all agents had exactly equal parameter settings. Trust update was set to 0.3 for positive and 1.0 for negative experience, honesty decay to 0.3, and punishment effect to 1.0. The confidence parameter was set to 0.95. The games were terminated after 200 transactions, or as soon as one of the producers ran out of stock.

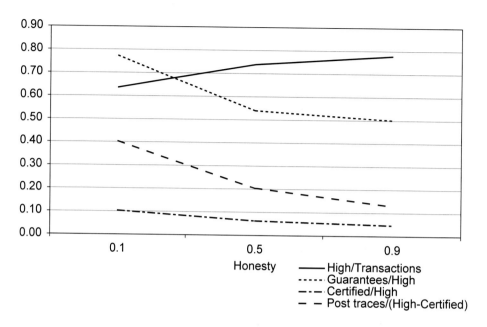

Figure 5.5. Sensitivity of trust update parameters.

In short games, decreasing initial trust is expected to decrease the proportion of high-quality transactions, to increase the number of certificates and guarantees, and to increase the tracing frequency. Indirectly, decreasing average trust is expected to decrease the deceit frequency in games with agents that are not perfectly honest, because a decreased proportion of high quality and increased certification decrease the opportunity to deceive and more intensive tracing will increase honesty through punishment.

Figure 5.6 shows that the sensitivity for initial trust is as expected (see Table A3 in Appendix F). The aggregated statistics of games with initial trust set to 1.0, 0.5, and 0 show a decreasing trend for the proportion of high-quality transactions (H/N=0.55, 0.47, and 0.43, respectively). The proportions of certificates (C/H) and guarantees (G/H) both show an increasing tendency. Tracing is increased (P/(H-C)=0.06, 0.37, 0.74) and deceit is decreased (D/N=0.05, 0.04, 0.02).

Sensitivity for honesty

The deceit frequency is expected to be strongly correlated with the honesty of the agents. The average level of honesty is also expected to have some indirect effects on game statistics. The average value of trust will decrease when tracing reveals the deceit. As a consequence, the proportion of high-quality products is expected to decrease as well, and the frequencies of certificates, guarantees, and tracing are expected to increase.

The organisation of transactions

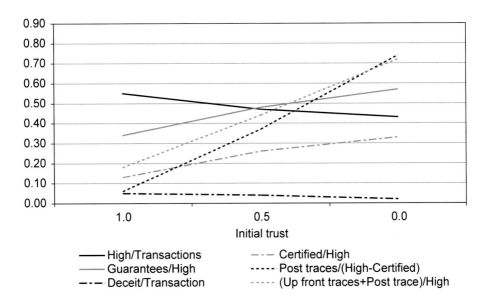

Figure 5.6. Sensitivity of initial trust.

The left-hand side of Figure 5.7 presents aggregated game statistics (see Table A4 of Appendix F) for games with different values of honesty, taken from Table A2 of Appendix F. Initial honesty and minimal honesty are set to equal values in these games. As expected, the right-hand side of Figure 5.7 shows that honesty has a strong effect on the deceit frequency. No deceit occurs in simulated games with initial and minimal honesty both set to 1.0. Setting of the honesty parameters has no effect in games with thrifty agents, because they offer little opportunity for deceit. The strongest effect of the settings of the honesty parameters is found in games with opportunistic agents. These games are the most sensitive for honesty, because opportunistic agents accept the risk of deceit for transactions with a good quality/price ratio. The tracing frequency depends on honesty as expected ($P/(H-C)$=0.31, 0.33, 0.37 respectively).

Figure 5.7 presents the effect of different values of honesty. Of course, simulated games with completely honest agents have the highest proportion of high-quality transactions, and only a small proportion of certified transactions. However, the frequency of high-quality transactions is higher than one would expect for the completely dishonest agents. This is to be explained as follows. In the beginning of the game, any tracing request will reveal deceit. Average trust is decreased and the average tracing frequency is increased, thus strongly reinforcing honesty. Some agents trace in the beginning of the game, some don't, for instance because the commodity was low quality or certified. The cause of the relatively high frequency of untraced high-quality transactions is the rapid decrease of trust in the first links of the network. A tracing request will always reveal deceit, reduce the remaining trust, and increase the tracing frequency in subsequent transactions. On the other hand, the punishment effect of tracing

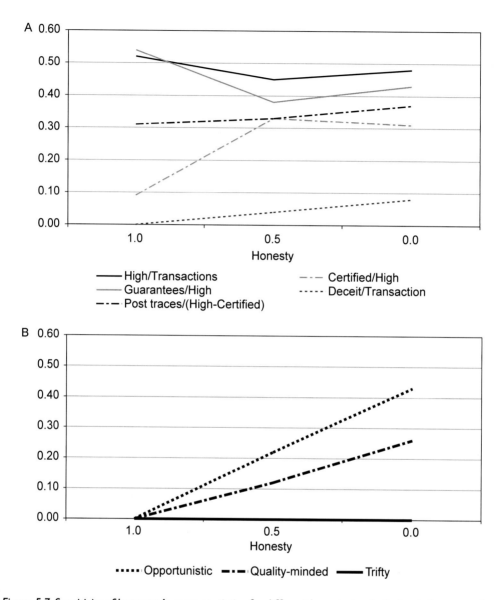

Figure 5.7. Sensitivity of honesty; A: game statistics for different honesty levels; B: deceit frequency for opportunistic, quality-minded and thrifty agent for different honesty levels.

increases average honesty more rapidly than trust decays: deceit reduces the trust of a single buyer, but all buyers benefit from the honesty that is reinforced by tracing. 'Knowing' that they will not dare to deceive, sellers offer guarantees for a very attractive price. Thus the delicate equilibrium between trust, tracing frequency and honesty is reached sooner in games that start from complete dishonesty than in 'half-honest' games.

The frequency of certificates (C/H) is much greater in games with dishonest agents. There is a shift from guarantees (G/H) towards certificates (C/H) if honesty decreases. The reason for this is that even after having given a guarantee, sellers may deliver untruthfully. The rapid decrease of trust in games where much is revealed also decreases trust in guarantees. The total frequency of certificates and guarantees increases with decreasing honesty, as expected (C/H+G/H= 0.61, 0.71, 0.74, respectively).

Endowment effect

The fourth series of simulation results demonstrate the sensitivity of the simulated game statistics for the so-called endowment effect: people experience loosing something they possessed as more painful than not gaining the same thing if they did not possess it. The endowment effect entails a high value of trust update after a negative experience and a low value of trust update after a positive experience.

We expect the endowment effect to lower the average trust level, so in games with endowment effect the proportion of high-quality transactions will be lower, the frequency of certificates and guarantees will be higher, and the tracing frequency will be higher than in games without endowment effect.

Simulation of the endowment effect requires the introduction of dishonest agents. Figure 5.8 (see also Table A5 of Appendix F) presents statistics of simulated games in which all agents have

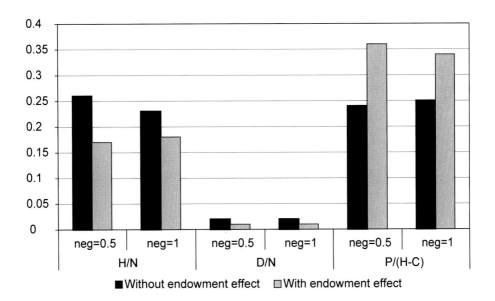

Figure 5.8. Sensitivity of endowment effect.

initial honesty = 0.5 and minimal honesty = 0.5. Other settings are unchanged with respect to the previous subsection, except for the update parameters represented in the Table A5.

As expected, the endowment effect is shown to decrease the proportion of high-quality transactions (H/N= 0.17, 0.18 with endowment effect versus 0.26, 0.23 without endowment effect). The cheating frequency (D/N) is lower in games with an endowment effect, because it takes a long time to heal the negative experience of punishment. Yet the endowment effect decreases the proportion of high-quality transactions, because average trust is lower. Once deceived it takes long to regain trust. Variance is high in games with a strong endowment effect or high sensitivity (δ^+=1 and δ^-=1), due to the fickle behaviour of the agents. Statistics of these games of a limited number of transactions are more sensitive to the coincidental revelation of deceit early in the game.

The impact of the endowment effect on certificates and guarantees is not that obvious as we expected (C/H+G/H=0.86, 0.76 with endowment effect, versus 0.74, 0.74 without endowment effect), probably because the number of certificates and guarantees is high anyway given the settings of the other parameters used in these games. Tracing frequency, as expected, is higher in games with endowment effect (P/(H-C)=0.36, 0.34 with endowment effect versus 0.24, 0.25 without endowment effect).

5.3.4.2 Testing hypotheses

This subsection offers a preliminary validation of tendencies reflected by the multi-agent simulation. We compare tendencies in multi-agent simulations with tendencies observed in human games, as formulated in the hypotheses in Subsection 5.3.1.

> *Hypothesis 1: When the initial willingness to trust is high the percentage of high-quality products sold will be higher than when the initial willingness to trust is low.*

Hypothesis 1 is directly confirmed by the data in Figure 5.5: the aggregated statistics of games with initial trust set to zero, 0.5, and 1.0 show an increasing trend for the proportion of high-quality transactions (H/N=0.43, 0.47, and 0.55, respectively).

For the purpose of testing the remaining hypotheses, the results from Table A2 in Appendix F are aggregated per strategy in Table A6.

> *Hypothesis 2: In a homogeneous environment with all opportunistic agents there are more cheats than with other profiles.*

Hypothesis 2 is confirmed by the results: D/N=0.07 for the opportunists, 0.04 for the quality-minded, and 0.00 for the thrifty agents, the latter simply giving little opportunity for deceit (Figure 5.8). Thrifty agents prefer low quality unless they can get certified or guaranteed high-quality products for a very good price. Quality-minded agents prefer to avoid the risk of deceit by trading certified products, thus reducing the possibility of deceit at the cost of up-front tracing, but unlike the thrifty agents they will buy high quality even if a risk remains. Opportunistic agents prefer high quality and in addition they prefer a good price over certainty. In the buyer role they take an increased risk of being deceived; in the seller role they are easily tempted to deceive. As a consequence deceit occurs most frequently in games with opportunists.

> *Hypothesis 3: In a homogeneous environment in which all agents are thrifty, i.e., who want to be certain about value for money, there will be more traces than with other profiles.*

Hypothesis 3 is confirmed by the results for (Q+P)/H in Figure 5.8 (i.e., the sum of up-front and post-transaction tracing requests relative to the number of high-quality transactions). The absolute number of traces is low in games with thrifty agents, because they trade few high-quality products (column H/N in Table A6). However, relative to the number of high-quality transactions, they have the highest tracing level. The tracing frequencies found in simulated games with quality-minded agents are not as high as for thrifty agents, but relatively high compared to the opportunists, due to up-front tracing in order to certify products before selling them. This increases price, but the quality-minded are willing to pay for the certainty that comes with it.

> *Hypothesis 4: Thrifty agents buy fewer high-quality products than opportunistic and quality-minded agents.*

Hypothesis 4 is confirmed by the detailed data in column (Q+P)/H of Table A7 of Appendix F (the aggregation level of Figure 5.9 does not give sufficient information to compare with Hypothesis 4). The proportion of high-quality transactions is less sensitive for trust in the quality-minded games than it is for the other strategies. The quality-minded prefer high quality and if they distrust they are likely to compensate the risk by either up-front or post-transaction tracing (high values of (Q+P)/H and P/(H-C) in the lower rows for quality-minded in Table A3). They are prepared to pay a higher price to reduce uncertainty.

Opportunists also trade many high-quality products, but they pay for it in a different way. They accept the risk of deceit more easily. This is an effect of the low weight of risk evaluation in their utility functions that makes opportunists negotiate about price or potential profits. The latter makes them trade more high-quality products. Opportunists agree to trade with a money-back guarantee (column G/H) more often than agents with other strategies, who prefer a certificate (column G/H). This is due to lower cost of guarantees with respect to certification and the low weight of risk evaluation in the negotiation models of opportunists.

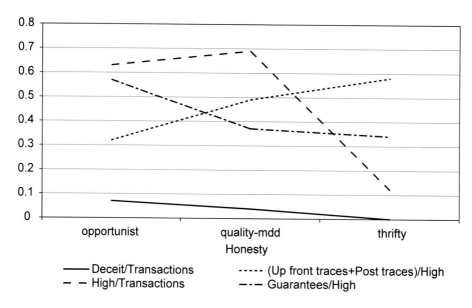

Figure 5.9. Sensitivity of strategies.

Agents following a thrifty negotiation strategy prefer to pay a low price, so they will not easily agree on buying with certificates or guarantees. In addition, they avoid the risk of being deceived. This results in a low ratio of high-quality products (H/N) being traded.

The simulation results presented so far concerned games with homogeneous agent populations. Hypothesis 5 is about games with differently configured agents in the same game.

> Hypothesis 5: In a mixed setting with opportunistic and thrifty agents, the opportunistic agents cheat less than in a mixed setting with opportunistic and quality-minded agents.

This hypothesis reflects observations of human simulation games that reveal extreme values of deceit frequency between games with mixed populations of opportunists and quality-minded on the one extreme and games with mixed populations of opportunists and thrifty agents at the other extreme. Figure 5.10 (see also Table A7 in Appendix F) presents statistics from simulated games populated with agents with different strategies. Some agents were configured to follow an opportunistic strategy. The others followed a quality-minded strategy in the first series of games, and a thrifty strategy in the second series of games. The simulation statistics confirm Hypothesis 5: much deceit in the first series (D=114), little deceit in the second series (D=23), although all agents are configured as dishonest. The numbers are the sum of all deceits in the three runs of the two series, where it is also the case that all individual runs of the first series resulted in more deceits than the second series.

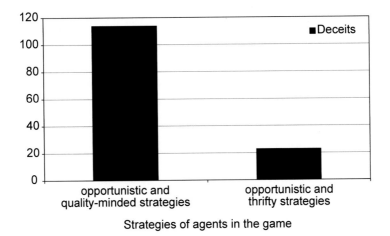

Figure 5.10. Deceits in games with mixed strategies.

The experiments confirm the validity of the models against the five hypotheses based on experience in human gaming simulations, formulated in Subsection 5.3.1. Thus one cycle of the research approach presented in Figure 3.6 was completed. So far we have only validated the tendencies, i.e. 'thrifty agents buy fewer high-quality commodities than other agents'. A quantitative calibration of the model would require additional cycles.

5.3.5 Discussion and conclusions

The research method for the development and validation of agent-based simulation of economic institutional forms consists of a cycle of steps, which can be started anywhere in the cycle depending on the state of data, theories, and hypotheses available. Gaming simulation is used to obtain reliable data about the economic institutional form being studied. The advantage of this approach is that different variable settings can be tested in experimental setups, something that is generally impossible to do in real situations. The data obtained are analyzed, leading to the conclusions about tendencies at the aggregated level. Theories are formed on the basis of those tendencies. The theories behind and design of the gaming simulation are translated into multi-agent models that are then incorporated into a computer simulation. Within the computer simulation the same variable settings can be loaded as were used in the gaming simulation. As a result the tendencies in the data obtained from the computer simulations can be compared to those obtained from sessions. Observed differences lead to adjustments of theories, with which the cycle can be repeated.

Sensitivity analysis by exploring variable settings is much faster with a multi-agent simulation than with gaming simulation, as groups of participants need to be found. Especially in a case like the Trust and Tracing Game where participants could only play once in a session the number of participants required increases with each variable to be explored. Therefore, one of the steps in the method is a systematic exploration of variable settings in the computer simulation, to a level of exhaustiveness as deemed necessary. The exploration is used to determine variable settings that are worthwhile to play out using human participants in a gaming simulation. The exploration of various variable settings is also of importance for verifying that the models correctly represent the theories about individual behaviour. For instance, it is possible to rapidly generate results for different institutional settings (size of the products supply, cheating fine, publication of tracing results, number of agents per role) and produce results similar to Figure 3-9 (Table A1-A7). Strategies of human game players can be analyzed by running agent simulation with various player strategies.

The approach is illustrated by applying it to commodity supply networks, and more specifically to the Trust and Tracing Game. This is a trade game on commodity supply networks, designed as a research tool for human behaviour with respect to trust and deceit in different institutional and cultural settings. Using the approach, individual-level agent models have been developed that simulate the trust, deception and negotiation behaviour of humans playing the Trust and Tracing Game. The models presented have been verified and validated by a series of experiments performed by the implemented simulation system, of which the outcomes are compared on the system level to the outcomes of games played by humans. The experiments cover in a systematic way the important variations in parameter settings possible in the Trust and Tracing Game and in the characteristics of the agents. The simulation results show the same tendencies of behaviour as the observed human games.

Aside from showing the validity of the research method introduced here, the paper also presents the agent models that simulate trust, deception, and negotiation behaviour of humans when playing the Trust and Tracing Game, with respect to a number of hypotheses and theories.

In an environment of perfectly honest traders, the speed with which agents learn that they are dealing with perfectly honest traders is directly related to the trust update parameters. The higher the trust update parameters, the higher the proportion of high-quality transactions, the lower the proportion of certificates and guarantees, and the lower the tracing frequency.

The endowment effect is the hypothesised effect that people experience losing something they possessed as more painful than not gaining the same thing if they did not possess it. As expected, including the endowment effect in the computer simulations is shown to decrease the proportion of high-quality transactions, lower the cheating frequency and increase the tracing frequency. Explanations are that in such games average trust is lower, and that once deceived it takes a long period to regain trust.

The research confirms the following hypotheses: there is more deceit in games with opportunistic agents than with other strategies, there is more tracing in games with thrifty agents, and games with quality-minded agents have lower tracing frequencies if trust is higher. During the validation phase also, two series of games were compared. In the first series some agents are opportunistic and others are quality-minded. The other series of games had opportunistic agents and thrifty agents. In all series agents were configured as dishonest. The level of deceit was higher in the first series than in the second series.

In current and future work more variations of the setting (including the current one) will be tested in both the human and simulated environment. This will lead either to further adjustments of the multi-agent model or to more variations to test. By testing large numbers of settings quickly in the simulated environment we can select more interesting settings for the human sessions, and thus save research time. The long-term result will, hopefully, be a fully validated model of trust with respect to situations comparable to the Trust and Tracing Game, where validation is reached for the individual and the aggregated level.

Finally, the research method introduced in this paper shows promise in terms of contributing a method from agent-based economics to new institutional economics by the use of human simulation games as intermediate steps. In human games, transactions can be monitored in much more detail than in real-life trade. That way it is much easier to collect the necessary data to set up dependable agent-based simulations of the situations enacted in the human games. The agent-based simulation will be informative for the more generic case in as far as the gaming simulation is a proper reflection of real-life trade processes. The results presented in this paper show that an agent-based simulation of the Trust and Tracing Game is indeed possible. The approach furthermore opens up the possibility of modelling specific groups of humans in trade situations (e.g. traders from different cultures). This will lead to agent-models that reflect the background of the group under consideration, and improve the accuracy of the models.

5.4 Validity

In chapter 2 the four criteria of Raser (1969) for the validity of gaming simulation as a research method have been introduced, namely: psychological reality, structural validity, process validity and predictive validity. Table 5.15 presents a matrix of how the 6 inputs to a session work out on the four criteria for validity. This table does not proof the validity of the TTG but shows the strengths and weaknesses as observed and experienced by the researchers. A discussion of this table follows in Chapter 7.

The multi-agent simulation (MAS) modelled after the TTG presented a whole new field of discussion considering reliability and validity. The MAS did not model the same reality as the TTG did, but it modelled the TTG sessions. The MAS had so many variables that simply calculating an equilibrium outcome given the parameter settings was not possible. While the multi-agent techniques used in this book are the contributions of the other authors of Chapter 5.3, it should be said in this book that the strict heuristic models that we started with during the MAS development could not be tuned to any outcome. The model has been tested for sensitivity on the major variables, and other variables have been set to 'neutral' or 'rational'

Table 5.15. Matrix of elements of the Trust and Tracing Game and criteria for validity.

	Psychological reality	Structural validity	Process validity	Predictive validity
Roles	Trader roles are individual roles, which made the performance of a role the performance of the person. This helped in getting involvement of participants.	Roles are abstract representation of real-world supply network roles. Consumer role is more like a large industrial end-consumer as they have a lot of money compared to the number of products and money at the traders.	Processes in TTG are limited to making transactions and tracing. No company-internal processes or transportation, etc.	Conclusions about individual traders in the session cannot be one-to-one translated to the real world. Focus on trade process.
Rules	Rules did not limit the imagination of participants to envision actions.	-	Only two rules: do not open envelopes and don't fight with each other. All other actions are allowed. This leaves the participants to shape their own process.	The two rules made that participants could enact all behaviour they would do in the real world. The sessions with a tomato supply network showed that real-world relations and behaviour was taken into the session.
Objectives	Money or points as a single objective. This created involvement and fanatic behaviour.	Building business relationships is not rewarded in the objectives, but could be a means to reach the goal.	Objectives made for a clear goal that started the trade process.	Single objective for participants does not immediately reward building trust and business relationships.
Constraints	-	Face-to-face negotiations, no other media for negotiation.	No constraints for behaviour within the face-to-face room setting.	-

Table 5.16. Continued.

	Psychological reality	Structural validity	Process validity	Predictive validity
Load	The game money was treated as real money. Envelopes as a model for food had the disadvantage that no real hazard to the consumers was present, except for fewer points.	Sealed envelopes as a model for food products with a hidden quality attribute was a good model according to the tomato supply network session participants.	The reference prices, tracing cost and start amounts of money have been tuned to make people discover cheaper tracing at the middlemen and to get the trade process going quickly.	Short duration of session (max 45 minutes) could give participants who invested time in trust and business relationships a disadvantage.
Situation	Mostly classroom setting: most participants were student. Involvement in a real trader role was difficult for participants, but most were highly involved in their imagined role.	-	Participants had different degrees of previous knowledge of each other between sessions. The difference in trade process this caused was the object of study.	In the hypothesis generation cycles the importance of real-world knowledge for behaviour in the session has been shown. The use of student in the empirical cycle sessions disconnects the sessions from existing real-world supply networks.

settings. The paper in Chapter 5.3 shows the tests for macro-level similarities in outcomes tested against hypotheses. The hypotheses were formulated based upon the observations in Chapter 5.1. The validity could be better checked against the quantitative results of Chapter 5.2 as the quantitative results are measured in real numbers, as are the MAS outcomes. This would require a new development phase however, costing another year at least, as the variables that have been measured in the TTG do not match exactly with the variables present in the MAS. Here the parallel development of the quantitative sessions and the MAS shows its disadvantages. The time available did not permit to do it differently.

6. The Mango Chain Game

The Mango Chain Game (MCG) is a gaming simulation developed to study bargaining power and revenue distribution in the Costa Rican mango supply network for export mangos. It studies the searching and bargaining sources of transaction costs. The organisation of transactions expresses itself in the use of short-term or long-term contracts. While initially it was hypothesised that long-term contracts would be used when the embeddedness was high, it turned out differently. The influence of local norms and values on the importance of good partnership was key.

In the MCG the searching and bargaining sources of transaction costs have been observed. The searching is expressed in the chapter as the selection of contract partners. The bargaining is analyzed in bargaining power and revenue distribution across the supply network.

This chapter is a (slightly) reworked version of the paper by Guillermo Zuniga-Arias, Sebastiaan Meijer, Ruerd Ruben and Gert Jan Hofstede, titled: Bargaining Power and Revenue Distribution in the Costa Rican Mango Supply Chain: A Gaming Simulation Approach with Local Producers, published in Journal on Chain and Network Science, 2007 (2), pp. 143-160.

6.1 Research issue

Of all major tropical fruits eaten freshly in Europe mango is one of the few that is grown by local producers instead of large multinational-owned farms. In Costa Rica the local producers of mango are generally poor, as the revenues they get for their mangoes are low compared to the prices paid in the foreign consumer markets. It is frequently argued that smallholders face major disadvantages with respect to bargaining and that the distribution of revenues is dictated by more powerful downstream agents. This paper investigates the bargaining power of members of the mango supply network and the revenue distribution across the network to derive policy implications for improving the bargaining position of the local mango producers.

The mango supply network in Costa Rica includes a variety of agents. Mango transactions differ in terms of volume, quality, price and delivery frequency and the produce can be sold either at the local market or at international markets. Relationships between producer associations, (local and international) traders, retailers and consumers are structured through a complex sequence of transactions. By gaining insight into how negotiation skills and relationships influence the governance structure used (measured via risk distribution and contract choice), strategic options for improving the smallholder position could be identified.

The scientific contribution of this paper is two-fold: it investigates bargaining power and revenue distribution in the Costa Rica mango network taking into account social variables in a way not described before, and to do so it uses gaming simulation as a research method. The use of gaming simulation as a data gathering tool for this type of research is new.

This article provides an analytical framework and an empirical assessment of the agency characteristics and perceptions that influence bargaining power and distribution of revenues. The analysis is conducted at two levels. First, the bargaining power of the participants is explained. Performance of contracts is measured in terms of revenues of the participants. Second, the outcomes are explained by relating agency characteristics to contractual attributes, using multiple regression analysis and logistic regression. This permits an understanding of underlying structural parameters and behavioural motives that explain bargaining power and revenue distribution.

The remainder of this article is structured as follows. First, it discusses the organisation of the mango supply network in Costa Rica, followed by the design of the Mango Chain Game that models crucial attributes of the mango network. Section 4 shows an analysis of the participants' bargaining power, paying attention to the underlying motives for increasing the revenue of the participants in a transaction. Finally, the usefulness of the research method used is discussed for analysing the mango supply network and some policy implications are suggested for improving the bargaining power of smallholders in the mango supply network in Costa Rica.

6.2 The mango supply network in Costa Rica

The main mango varieties grown in Costa Rica and suitable for export are Tommy Atkins, Kent, Keith, Palmer and Smith (Mora, pers. comm., 2004; Jiménez, pers. comm., 2003; Mora Montero, 2002). In 2006, mango production was 14000 metric tons. 11000 metric tons were sent to the export market, and 3000 metric tons sold in CENADA (wholesaler market). During the years 1996 to 2003 the planted area increased by 7% and the quantity produced rose by 75% (Jiménez, pers. comm., 2003). The yield increase was mainly due to technological progress (Mora Montero, 2002). Costa Rica has about 1,950 mango producers of which 60% cultivate less than 5 ha, 35% cultivate between 5 and 20 ha and 5% own more than 20ha (Mora, pers. comm., 2004; SEPSA, 2001). The producers are organised in different ways; large and medium-size producers are linked to international trading companies and small producers are either independent or affiliated to co-operatives or producers associations.

The Costa Rican Ministry of Agriculture provided access to reports (i.e. Mora Montero, 2002) and personal communication to collect data for describing the structure of the mango supply network in Costa Rica, distinguishing between production for the local and for the export market. Millet (2003) conducted a pre-study for this paper. The organisation of the mango supply network is relatively simple for the export market. For the local market the network is more complex (Figure 6.1) as there are a wide number of different intermediaries involved. The intermediaries play multiple roles in the network: they buy mangoes from the producers, sell the produce to the CENADA wholesale market, or buy from CENADA to deliver to local outlets such as local retailers and the wet market (about 20% of the total production). The mango in CENADA is sold to different agents that deliver the mango to other outlets.

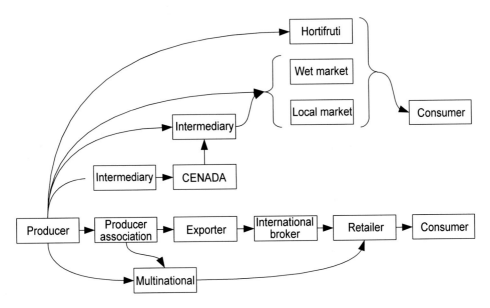

Figure 6.1. Mango supply network in Costa Rica.

Consumers preferences toward markets show that in 2006 38.6% prefer to buy in the Feria (HortiFruti), 12.9% in the local vegetable store, 28.5% at supermarkets, 10% at the municipal market and 10% in other outlet (CNP website, 2007).

A governance structure consists of a collection of rules, institutions and constraints structuring the transactions between the various stakeholders (Hendrikse, 2003). There are three governance structures: spot market, hierarchy (vertical integration) and network.

In Costa Rica the export market is governed by contracts with network and hierarchy (vertical integration) characteristics, whereas in the local market spot market is the main governance structure. Vertical integration is present in the export market where the multinational company controls many of the actions of the producer associations. The necessity of meeting international quality standards such as EUREPGAP increases the presence and control of exporters and retailers on the actions of the mango producers in Costa Rica. In addition, there are some examples of network forms of governance, such as relational contracts which are contracts that are not governed by (written or oral) contracts, but by maintaining a good relationship.

6.3 Analyzing bargaining power in the supply network

Muthoo (2000, 2002) defines a *bargaining situation* as a setting where individuals engage in mutually beneficial exchange but maintain conflicting interests over the terms of the trade.

Muthoo (2000) explains *bargaining* as any process through which the agents try to reach an agreement. This process is typically time-consuming and requires agents to make offers and counter-offers to each other.

Chamberlain and Kuhn (1965) define *bargaining power* as the ability to secure an agreement on one's own terms. In Leap and Grigsby (1986) Chamberlain (re-)defines bargaining power as the cost of agreeing (or disagreeing) with the opponent. A party's bargaining power increases as the cost of disagreeing with an opponent decreases. In a similar vein, Slichter (1940) defined bargaining power as the cost to one agent of imposing a loss upon another agent. The notion of bargaining power is rooted in power-dependency theory (Emerson, 1962), which states that one agent's bargaining power is derived from another's dependency. More specifically, Agent A's dependence on Agent B is directly proportional to A's motivational investment in goals mediated by B, and inversely proportional to the availability of those goals to A outside of the A-B relation. Therefore, an agent who can develop or gain bargaining power is able to reduce dependency on other agents (Cook, 1977; Bacharach and Lawler, 1984 as quoted by Yan and Gray, 2001).

6.3.1 Determinants of bargaining power

The major determinants of bargaining power that will be exposed here are (1) wealth, (2) skills of the negotiator, (3) partnership and (4) market imperfections. Muthoo (2002) stresses wealth as a key determinant for bargaining power. The agent who perceives himself as wealthier will hold more bargaining power during negotiations, while poor agents will not be able to exert bargaining power. Following the same line, Fossum (1982) recognises two separate aspects of bargaining power: the power inherent to the economic positions of the parties (wealth) and the attributes and skills of the negotiator (individual characteristics). Lee *et al.* (1998) performed a literature review on international joint ventures and found that their bargaining power increased with the strategic importance of partnerships (Yan and Gray, 2001), with resources linkage between partners (Lecraw, 1984; Kumar and Seth, 1998), and with the availability of alternatives available to the partner firm (Yan and Gray, 2001). According to Leap and Grigsby (1986) this points to some key factors that affect bargaining power, such as availability and control of resources, potential and enacted power relationships, and absolute, relative and total power. Rubinstein and Wolinsky (1985) and Leap and Grigsby (1986) emphasised that the bargaining positions, and hence the agreement reached in any particular setting, will be affected by the conditions prevailing at the market (market imperfections). These include the opportunities that each of the negotiation parties have for meeting other partners in the event that the agreement in the current negotiations is delayed, including the expected length of time required to achieve any alternative transaction and the expected behaviour of alternative partners.

6.3.2 Determinants of revenue distribution

Revenue distribution is the accumulated end result of all transactions in the supply network. Depending on the field of interest, researchers can try to relate many aspects to revenue distribution. Cultural anthropologists will look at the cultural backgrounds; engineers will look at the technical facilities of the companies, etc. This paper focuses on the aspects that are relevant for explaining governance structures, because a governance structure affects the size of the surplus that will be generated, by its effect on investment, efficiency of bargaining, and risk-aversion (Zingales, 1998, quoted by Hendrikse, 2003). Here particular attention is paid to the bargaining and risk-aversion. Influence on investments (in quality) is the topic of a separate paper (Zuniga-Arias, 2007).

For the analysis of revenue distribution bargaining and risk aversion are disentangled into four independent variables: contract choice, bounded rationality (information problems), bargaining power and risk perception.

Contract choice

Contracts reached during the bargaining process provide a mechanism for sharing or avoiding risk and for reducing the uncertainties produced by bounded rationality, and tend to reflect the (perceived) bargaining power. An important aspect of supply network transactions is related to the distribution of bargaining power amongst agents (Nash, 1950; Rubinstein 1982). The choice of potential trade partners is therefore influenced by the anticipation of their behaviour and response.

Bounded rationality

Close agency relationships are a vital element for supply network integration, but industrial experience has shown that it is difficult to contract across inter-organisational borders. Agents will only accept such contracts if they are willing to share both risk and rewards amongst members of the supply network (Agrell *et al.*, 2002). Such willingness runs counter to the instincts of most entrepreneurs in most places in the world (Hofstede, 2004). Contracts can be considered as mechanisms to share the risk and reduce the uncertainties of the transactions. Asymmetric information in supply networks can be used by agents to reduce their own risk and costs to the detriment of the overall supply network performance. Agrell *et al.* (2002) show the close interaction between risk-sharing, contracts and information asymmetries (*information problems*) between agents in the network. Reallocation of bargaining power does not necessarily lead to better performance and improved coordination in the supply network. There is little or no evidence either that contracts are preferred in situations where agents are more risk-averse (Roumasset, 1995).

Bargaining power

Regarding the effect of the contextual conditions on contract formation, Argyris and Liebeskind (1999) and Stinchcombe (1985) argue that the contracting agents' ability to influence the terms and conditions of contracts is highly contingent on their bargaining power (Buvik and Reve, 2002). If there are fewer buyers than sellers, this will increase the buyers' bargaining power vis-à-vis the suppliers (Heide and John, 1992). Higher bargaining power can increase stakeholders' marginal revenues and hence their provision of effort (Chemla, 2005). Taking into account that the revenue is the total income produced by one agent, one can assume that the more revenue received in a specific transaction, the more bargaining power the agent has, considering the income that a particular stakeholder receives from a negotiation as an expression of bargaining power (Dobbelaere, 2003).

Risk perception

Risk is an important topic in contract design because it affects both the cost of risk-bearing and the motivation to behave in a certain desired way (Bogetoft and Olesen, 2004). According to Fafchamps (2004), comparing agricultural case studies from Madagascar, Benin and Malawi, if a producer is a risk taker then the effect on the revenues is likely to be positive. Muthoo (2002) stresses that asymmetric information reduces bargaining power, and Ayala (1999) points out that less available information increases the uncertainty and risks of making wrong decisions. Because asymmetric information is also a cause of bounded rationality (see: Fafchamps, 2004) it is likely that agents with more information ('more relaxed bound on rationality') will receive higher revenue shares.

6.3.3 Analytical model

For this study an analytical model (Figure 6.2) has been developed to structure the dependent and independent variables. The final dependent variable represents the revenue distribution between buyers and sellers. Independent variables for the revenue distribution come from the theory reviewed above (Risk perception, Contract choice, Bounded rationality and Bargaining power). Bargaining power is treated as a special independent variable as it is the dependent variable of the first list with independent variables, identified in the first subsection of this section (Wealth, Partnership, Market imperfections and Negotiator skills).

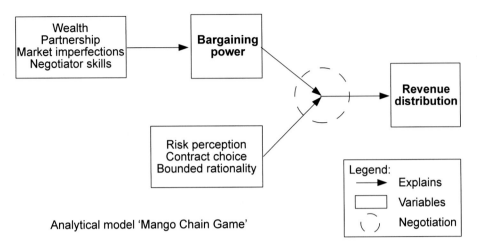

Figure 6.2. Analytical model of bargaining power and revenue distribution in MCG.

6.4 Materials and methods: the Mango Chain Game

The design cycle of the Mango Chain Game (completed in October 2004) consisted of a pilot development and testing of the gaming simulation in the Netherlands and its fine-tuning in Costa Rica. Sterrenburg and Zuniga (2004) describe the Dutch development phase in detail, while Meijer, Zuniga and Sterrenburg (2005) outline the Costa Rican development and testing phase.

During the empirical cycle of the research method 5 data-gathering sessions were performed. Based on the case description, on theory and on induced observations from the test sessions, an experimental session set-up was constructed that consisted of the load (the configuration of variables of a gaming simulation in a particular session) (Table 6.1) and situation (the choice of participants, venue, measurement techniques, introduction and facilitation) (Table 6.2).

6.4.1 Description of the Mango Chain Game

The Mango Chain Game is played on a game board placed on a large table, with enough room for all participants to walk around. The game board reflects the structure of the mango network (Network in Figure 6.1, four boards in action in Figure 6.3). Products traded are coloured fiches representing mangoes of different qualities. The Mango Chain Game defines four specific agents with their respective roles:

Producers associations' goal is to earn as much money as possible. By randomly selecting a coloured fiche from a dark box they are informed about the volume of production from their producers. Producer associations can influence the quality level of their supply by investing a

Table 6.1. Experimental session setup (load).

Topic	Configuration
Configuration of roles	3 or 4 producer associations (1 person), 1 - 2 independent exporters (1 person), 1 multinational company (2 - 3 people), 1 - 2 retailers (1 person).
Tools per role	4 production fields per producer association, 2 premium, 2 normal quality. One boat per independent exporter / multinational, one truck per producer association. Boats for rent for retailers.
Start amounts (money)	Producer associations: 200; independent exporters: 480; multinational company: 760; retailers: 730.
Consumer market	Prices determined by the market, when prices go up then supply goes down and vice versa.

Table 6.2. Experimental session setup (situation).

Topic	Configuration
Participants per session	All members of one producer association, all active in the association either as producer, worker for a producer or as association employee.
Session duration	2.5 to 3 hours of actual playing time.
Measurement of bargaining power and participant properties	Via post-session questionnaire.
Measurement of transactions	Via contract form, recorded for every transaction in the simulation, following the main three governance structures spot market, network and hierarchy (vertical integration).
Goal of observations	Provide qualitative additional comments to quantitative measurements.
Observation method	2 observers plus observations from 2 game leaders.
Introduction method	Plenary explanation followed by a 'round 0': a game round that will not have any consequences for the outcomes as the actions will be undone after the round ends.
Facilitation (game leaders)	2-3, they control the behaviour of the session, one will play the producer market (basically production handling, natural hazards, time control), another the consumer market (transport risk, manage the selling process from the retailers), and the third one the bank role (contract check, rent of boats, observer).

The organisation of transactions

fixed amount of money per production field. They are the only agent able to sell in the local market.

The multinational represents a dominant agent in the supply network of mango. This is the only role that can be played by more than one person. The initial operating capital of the multinational company is large. The goal of the multinational is to get as much revenue as possible. The multinational buys from producer organisations and sells to retailers.

Independent exporters are an alternative for the multinational. They have the same goal, but are considerably smaller which is expressed in only one participant per exporter and a more limited operational capital.

Retailers have exclusive access to the consumer market. The goal of retailers is revenue maximisation too. They have to deal with demand uncertainty in the consumer market.

6.4.2 Mango chain game process description

The MCG is played in several rounds, each consisting of the following steps:
- **Production**
  Production is simulated by game leaders. Each of the production fields produces two tons of mango (represented by dry beans in various colours) per round in the quality determined by the quality of the production field. The amount of mango per field has a bandwidth of 50%: by picking a coloured fiche at random from a dark box the producer organisation has ¼ chance of 50% less production, ½ chance of normal production and ¼ chance of 50% more production. The mangoes (beans) are placed in a cup of the producers association.
- **Trade**
  During the trade phase participants walk around through the room and try to make contracts with each other. Contracts are signed on pre-structured forms with variables duration, price, quantity, quality and allocation of risk. Contract choice is determined by the duration of the contract (see appendix 2 for the contract). Risk allocation refers to three types of uncertainties present in the gaming simulation; (1) variability in the supply from the simulated producers, (2) quality loss with a chance of 1/6 in each transport stage, and (3) uncertainty about the consumer market price. The trade phase ends with the handing over of all signed contracts to the game leaders who will check them for accuracy and then place a coloured elastic band between the traders on the game board. The colour of the elastic band represents the duration of the contract.
- **Transportation**
  During transportation there is a chance that quality decreases. Perished goods are destroyed immediately at no cost. For both transport modes (trucks and boats) a dice is thrown. All trucks and boats experience the same good or bad luck at once. This represents real-world uncertainty such as transportation damage, storage capabilities, delays or rough handling.

In a similar vein, goods that are left over after a round decrease one degree in quality, as they get older.

- **Contract fulfilment**
  Once the sellers know what quantities of mangoes are left after transportation they can decide which client will get which mangoes, in case there are more contracts. In the case of a shortage of mangoes, care must be taken regarding opportunities for contract breach. Payments for supplies are done at once upon delivery.

Consumer market

Similar to the production side, consumer demand is game-leader simulated. Retailers inform the game leader how much they want to sell to the consumer market. The game leader then determines prices per unit based upon the total supply. High and low quality has a different price. The domestic market buys all left-overs from producer associations, multinational and independent exporters at a fixed low price regardless of quality.

6.4.3 Conduction of the data sessions

Costa Rican mango producers participated as agents in the five sessions conducted. For each session, the researchers first visited the local producer association, to assure that the local leader of the association could mobilise sufficient participants. Most of the participants owned a small- to medium-sized mango plot. Some were employee of a medium-sized mango plot or of the association itself. Role assignment to participants in a session was random.

6.5 Operationalisation of the analytical model

The analysis was based on two data sets collected during and after the game sessions. The transaction forms used in the sessions have been collected to register the contract choice and the different exchange attributes (price, volume and quality). After the gaming simulation, all participants filled in a questionnaire about risk attitudes, access to information, perceptions of market access, and determinants of bargaining power. The transaction data set constituted 82 records (contracts) between a seller and a buyer. Forty-three questionnaires have been collected. To determine the partition of the revenues both samples have been merged.

6.5.1 Bargaining power

In the analytical model bargaining power depended on wealth, partnership, negotiation skills and market imperfections. *Wealth* was related to tangible and intangibles assets in the gaming simulation. The tangible assets differed per role. Wealth was measured as self-perception of the wealth exercised during the session in the questionnaire. *Negotiation skills* were related to the background of the participant. The questionnaire asks a participant to give a subjective measurement of his capacity to negotiate prices. *Partnership* was related to cooperation and building of trust and friendship. Participants provide a subjective valuation of the main business partner in the session. Some are friends and some will like each other in real life. These relationships participants take with them into a session. *Market imperfections* related to uncertainty and asymmetric information. In the gaming simulation configuration there is much uncertainty that affects agents in a different way. The questionnaire asked how much market imperfections influenced the bargaining power.

Following Muthoo (2002), all variables were measured on a Likert scale ranging from 0 (low) to 10 (high). Descriptive statistics are reported in Table 6.3. The results show that multinationals scored the highest on every aspect. Producers have relatively low wealth, but have the second-highest bargaining power. This is most probably due to their hold-up capacity. Multinationals and exporters are strongly dependent on partnerships to guarantee continuous sourcing. The retailers scored lowest on bargaining power, negotiation skills and partnership. This could be the effect of all risks present in the network. All agents are affected by market imperfections.

Table 6.3. Descriptive statistics of the agency attributes.

	Producer associations		Multinationals		Independent exporters		Retailers	
	Mean	S.D.	Mean	S.D.	Mean	S.D.	Mean	S.D.
Bargaining power	7.3	2.6	**8.2**	1.2	6.6	1.3	**5.9**	1.6
Wealth	**5.4**	3.0	**7.8**	1.9	5.7	3.1	5.7	2.3
Negotiation skills	7.1	2.0	**9.2**	0.8	7.4	2.2	**6.7**	2.3
Partnership	7.9	2.5	**9.0**	0.9	8.1	1.5	**6.7**	2.8
Market imperfections	8.4	1.8	**8.7**	0.9	**8.1**	0.9	8.4	1.4

Note: values measured with Likert scale (from 0 to 10). The highest and lowest marks are labelled bold.

6.5.2 Risk perception

In addition to the agency attributes, all participants in the game were asked to answer some questions regarding their risk perception. Two methods for risk assessment were used: (1) a direct approach where participants were asked to define their risk attitude (risk averse, risk neutral, risk-taker) before the simulation game; and (2) an economic experiment[1] for risk measurement where the participants were asked to select amongst different lottery choices. Cross-tabulation and Chi-square test were performed to identify significant association between these variables (Table 6.4). After verifying the independence of both risk indicators, a principal component was performed in order to construct an index of risk perceptions.

Table 6.4. Relationship between risk perception and the risk experiment.

	Producer associations		Multinationals		Independent exporters		Retailers	
	Avg.	S.D.	Avg.	S.D.	Avg.	S.D.	Avg.	S.D.
Risk attitude[a,c]	2.0	2.27	2.2	0.49	2.3	0.38	2.5	0.33
Risk attitude[b,c]	1.36	0.13	1.60	0.25	1.75	0.25	1.38	0.18

[a]Scale from 1 to 3, where 3 is risk taker
[b]Scale from 1 to 2, where 2 is risk taker
[c]Pearson Chi-Square is significant, therefore a relationship exists between these variables

6.5.3 Contract choice

The contract forms permit three types of contracts: one, three or infinite round length.

Figure 6.3 shows an example of contract formation during one session. The round number is in the upper right corner; each rectangle represents the game board with 4 producer associations (PA) 2 independent exporters (IE), 1 multinational company (MN), and 3 retailers (R). (In the specific case of the session shown on the game board there were no participants for exporter 2 and retailer 3, so they have not been used). The rectangle on the left is the production market and the one on the right the consumer market; the rectangles at the top and bottom left are the local market. The triangles represent the production hazards and transportation hazards. The smiley face represents a good production season; a lightning arrow represents a bad one. The thick line represents a one-round contract and the thin line represents a three-round contract. As drawn in round zero, most participants started to develop trade relationships and used a short-term contract. Some participants faced transportation problems (marked with a lightning arrow too) (with a contract breach resulting from that). One transaction had been built with a three-round contract. In round one, repeated contracts appeared (contracts

---

[1] For more information about the economic experiment see Davis and Holt (1993). Chapter 8. Individual decisions in risky situations.

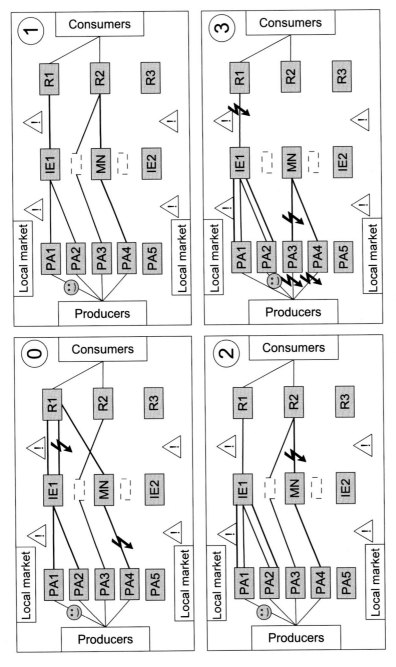

Figure 6.3. Game boards with contract configuration in one session (four rounds).

with the same trade agent as the round before), and some of them have been converted into a three-round contract. In round two and three it is possible to observe how some agents stick to short-term contracts with the same traders as the round before. (MN+ R2, IE1+ R1, PA4+MN) Effectively these contracts work like a long-term contract.

6.5.4 Bounded rationality

Agents making a contract each have partial information only. A proxy variable has been constructed for the 'bounded rationality' faced by agents during a session. The post-session questionnaire asked all agents to identify the different types of problems faced for getting an agreement. In order that the answers could be used they had to be merged with the information gathered from the transaction form (the transaction form holds the contract choice and the buyer and seller among other attributes). The answers were classified into problems related to information and other problems (Table 6.5). Then the ratio of information problems, other problems and no problems has been calculated to make the sum equal to 1 per role. This ratio was used as an independent variable in subsequent regressions as proxy for bounded rationality. Table 6.5 shows that multinationals had less information problems than the other roles, but more of the 'other problems'. There were no retailers that reported no problems.

Table 6.5. Distribution of the problems faced by the participant of the sessions.

	Producer association	Multinational	Independent exporter	Retailers	Average
Information problems	0.588	0.500	0.563	0.595	0.575
Other problem	0.353	0.438	0.375	0.405	0.383
No problem	0.059	0.063	0.063	0.000	0.042

6.5.5 Revenue distribution

Revenue was the amount of income earned per agent counted and recorded at the end of a session. Since each role had a different amount of start money the margins above the initial amount of money have been used as a performance indicator. Figure 6.4 shows the selling/buying price and the distribution of the revenues. In addition, Figure 6.4 reports the margins received for the different qualities and networks that were used. The figure starts in the producer associations' box. There are two prices here: 15 and 10. Price 15 is the bottom price for the premium quality and 10 the bottom price for the normal quality product, PA's on average sold their product at different prices depending on the actor they faced, for example, producers received better prices when they engaged in business with the multinational (25.6

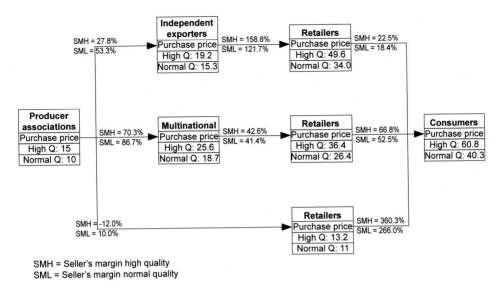

SMH = Seller's margin high quality
SML = Seller's margin normal quality

Figure 6.4. Revenue distribution in the different network configurations.

for the premium and 18. 7 for the normal). When multinational (MN) and independent exporters (IE's) sell the product to the retailers (R) the independent exporter gets the better price and margin. When retailers sell to consumers, R's get better margins when selling the product they bought from the multinational and from the producer associations.

6.6 Results

The results consist of two subsections, following the structure of the analytical model (Figure 6.2). The first subsection presents the results of the bargaining power and the main influencing attributes. The second subsection explains the revenue distribution and how it is influenced by the attributes in the analytical model. The data analysis consisted of three stages. Multiple regression models have been estimated for the determinants of bargaining power related to the agency attributes. Hereafter, logistic regression techniques were used to disentangle the differences in bargaining power between buyers and sellers, relying on the attributes of the contract partners as explanatory variables. Finally, the index of bargaining power is used as an independent variable for explaining the revenue distribution in the Mango Chain Game.

6.6.1 Bargaining power

It was determined how the attributes of the analytical model influenced the bargaining power (BP) of the participants. Results are reported in Table 6.6. Variables reflecting negotiation skills, partnership and wealth are all significant and exhibit a positive effect on the bargaining power of the participants. These three variables together explain 56% of the variance. Bargaining

Table 6.6. Determinants of bargaining power.

	Coefficient	Significance
Constant	-1.968	0.847
Market imperfections	2.958	0.811
Negotiator skills	0.325	0.017**
Partnership	0.375	0.001***
Wealth	0.194	0.055*

Adj. R^2 = 0.558; note: * sig. at 10%, ** sig. at 5%, *** sig. at 1%

power proved to be independent of market imperfections. For strengthening the bargaining power one could increase the skills of the people, to motivate long-term and harmonious relationships, and to increase agency wealth.

6.6.2 Revenue distribution

This subsection identifies the factors that explain the distribution of revenues between buyers and sellers. To be able to perform this analysis, two different analytical models were built, one for the buyer's revenue and one for the seller's revenue. The seller's revenue is defined as the purchase price of the good minus the selling price of the good. The buyer's revenue is defined as the expected selling price of the goods (determined by the mean selling price for the particular route in the network, see Figure 6.4) minus the purchase price of the goods (determined by the transaction with the seller). The bargaining power obtained in the first analysis for sellers and buyers has been disaggregated depending on their contractual arrangements. The information problems related to uncertainties faced during the session are specified too for both sellers and buyers. Similarly, risk attitudes of buyers and sellers have been defined using risk tests in the post-session questionnaire.

The variable *contract choice* is a binomial choice, since the participants only signed one-round or three-round contracts on the contract form.[2] The determinants of the buyer's revenue are reported in Table 6.7 and explain 47% of the variance. Factors negatively influencing the buyer's revenue are the bargaining power and perceived risk of the other agent. However, neither the seller's bounded rationality nor contract choice had a significant impact on the revenue distribution. Consequently, with a lower bargaining power and a higher risk aversion of the seller, a buyer may expect to receive higher revenues.

---

[2] The 'infinite' option has never been used.

Table 6.7. Determinant of the buyer's revenue (N=82).

	Coefficient	Significance
Bargaining Power Seller (BPS)	-0.775	0.048**
Perceived Risk Seller (PRS)	-1.321	0.0001***
Bounded Rationality (BRS)	-0.670	0.380
Contract choice	-0.317	0.108

Adj. R^2 = 0.466; note: * sig. at 10%, ** sig. at 5%, *** sig. at 1%

The determinants of the seller's revenue are reported in Table 6.8 and explain 35% of the variance. The bargaining power and risk attitude of the buyer and the contract choice negatively influence the seller's revenue. The bounded rationality of the buyer is not significant. The main difference with the factors influencing the buyer's revenue is that the contract arrangement has a significant impact on the seller's revenue. Since there were only two types of contracts (one round and three rounds), the negative sign of the coefficient means that the long-term contract decreases the revenues of the seller.

For both buyer and seller, the other party's bargaining power has a negative influence on revenue. This implies that powerful traders earn more at the expense of the less powerful trade partner. This is supported by Muthoo (2001). In a similar line, the risk perception of the trade partner has a negative influence on the revenue. This means that the Mango Chain Game makes risk-averse buyers willing to pay less for the mangoes if they have to take some of the risk present. Risk-averse sellers ask a higher price to be willing to take the risk. Again, this is in line with Muthoo (2002). The contract choice is only significant for the seller's revenue. As the contract choice is one round or three rounds only, the seller will get higher revenues when signing one-round contracts.

Table 6.8. Determinants of the seller's revenue (N=82).

	Coefficient	Significance
Bargaining Power Buyer (BPB)	-0.379	0.039**
Perceived Risk Buyer (PRB)	-2.025	0.027**
Bounded Rationality (BRB)	-1.057	0.236
Contract choice	-0.264	0.043**

Adj. R^2 = 0.352; note: * sig. at 10%, ** sig. at 5%, *** sig. at 1%

6.7 Discussion and conclusions

This article described a gaming simulation approach for analysing bargaining power and revenue distribution. The approach used contributes to the gaming simulation methodology, as the use of data from a gaming simulation on face-to-face trade is new. In the opinion of the authors, the main advantage of gaming simulation is that the results can be used to identify implicit issues involved in transactions that often escape notice in academic studies. In this study it has been possible to study multiple networks in the five sessions as one data set. Before this, however, a lot of iterative prototyping and testing was needed. During this phase the intermediate results showed that the outcomes of a gaming simulation are highly dependent on the rules, roles and incentives. A gaming simulation is a tailor-made precision tool that requires in-depth analysis of the research question and the possible factors of influence and a thorough fine-tuning before it can become a useful tool to answer research questions. Testing over and over again led to continuous improvement of the Mango Chain Game. About four times more test sessions than data sessions have been conducted.

Multinational companies obtained the highest bargaining power in the Mango Chain Game, followed by independent exporters. Retailers had the lowest power. In this sense, the gaming simulation was realistic in recreating the real-world mango network found in the study of Millet (2003). The producer association in the gaming simulation showed a relatively strong bargaining power compared to the information the participants gave during the sessions. This might be due to the modelling that left out the larger growers, thus leaving the producer associations all of the supply market.

In the analysis, characteristics explain revenue for both buyer and seller. As expected, a lower bargaining power of the buyer (seller) resulted in higher revenue for the seller (buyer). In general, stakeholders with more bargaining power have been able to take advantage of the other agents. Higher risk-aversion of the buyers and/or the sellers has led to more revenues for the other agents involved in the exchange relationship. In the same vein, long-term contract in the buyer-seller relationship have led to lower revenues (but also reduced risk) for sellers.

The latter result is surprising, since contract choice appeared to be only significant for the seller's revenue equation and not for the buyer's. Mango producers are well aware that the type of markets in which they operate are mainly based on short-terms contracts. This gives them the opportunity to remain flexible towards the changes in demand and supply that they cannot control. Otherwise, producers are also trying to establish long-term relationships, but they could equally rely on repeated short-term contracts with the same partner. The latter type of contracts tends to rely on trust or friendship. Here the network governance mechanism is at play.

An incentive to take into account in the transaction between buyers and sellers is that the latter prefer to rely on short-term contracts with the buyer that include not only price, delivery time and quality attributes but also stipulate risk-sharing possibilities to enable them to redress the risk averseness of the seller to the detriment of buyers' revenue. It is rational to believe that buyers do not want to change the status quo and are therefore not interested in putting their revenue share at risk. Mango producers will try to increase their bargaining power by improving their wealth, searching for good partnerships and increasing their individual negotiation abilities. In Costa Rica, this might be achieved by offering training in negotiation skills delivered by the National Learning Institute (INA), whereas the search for partners might be done in conjunction with International Trade Promoter (PROCOMER).

Finally, wealth appeared to have a significant impact on bargaining power. In terms of tangible assets, the government could establish credit facilities for small and medium-size producers that might be conditional on membership of a producer organisation. This will certainly increase the producers' bargaining power, since - as was observed in the gaming simulation - access to credit, resources and cooperation were key factors in the outcome of the negotiation process.

6.8 Validity

In chapter 2 the four criteria of Raser (1969) for the validity of gaming simulation as a research method have been discussed, being: psychological reality, structural validity, process validity and predictive validity. Table 6.9 presents a matrix of how the 6 inputs to a session work out on the four criteria for validity. This table does not proof the validity of the MCG but shows the strengths and weaknesses as observed and experienced by the researchers. A discussion of this table follows in Chapter 7.

Table 6.9 shows that modelling the Mango Chain Game after the results from Millet (2003) increased in particular the structural validity, and to some extent the psychological reality and process validity. Playing with real smallholders (situation) was positive for the psychological reality and process and structural validity. For the predictive validity this meant however that the results stem from the perspectives of smallholders only. The predictive reality however shows from the similarity between the real-world revenue distribution and the one obtained in the sessions.

Table 6.9. Matrix of elements of the Mango Chain Game and criteria for validity.

	Psychological reality	Structural validity	Process validity	Predictive validity
Roles	Roles similar to real-world chain. Participants knew what these roles meant in the real world.	Roles correspond with real-world roles in export network. Local market has been simplified. Positively appraised by participants.	Functions for each role w.r.t. trade are modelled after real-world process.	The low revenues that smallholders receive in the real world occurred in the sessions too. Bargaining power distribution similar to real world according to participants.
Rules	Simple rules: just enact your role while making contracts, but stick to time available in a round. Participants tended to continue negotiating between rounds, which showed their involvement. The rounds took them out of their involvement in actions.	Contracts based upon theory on aspects one can negotiate about. (Price, volume, quality, risk allocation).	Rounds structured the contract negotiation and contract fulfilment phases. Contract fulfilment is stylised on game board. Contract negotiation is free-form. Perishable product pushes willingness to make contract or sell on local market.	-
Objectives	Participants had to make contracts to trade and earn money. This is the normal behaviour of the participants.	Perishable product has to be sold is similar to real mango product.	To earn money while building business relations is similar to real-world objective.	It is the goal of the real-world producer associations to earn more money in the near future to gain more wealth for the smallholders.
Constraints	-	Expensiveness of bypassing multinational or independent exporter models impossibilities to deal directly with (Western) retail company.	No increase possible in production area, only in quality of area. This is a long-term option that would transcend the time-scope of the session.	-

The organisation of transactions

Table 6.9. Continued.

	Psychological reality	Structural validity	Process validity	Predictive validity
Load	Game board enhanced visualisation of whole network, unlike real-world information.	No alternative multinationals, relatively few producer associations per independent exporter / multinational	-	Size of risk has been tuned to real-world risk to make participants enact the same risk avoidance.
Situation	Played in villages with local smallholders. (Foreign) University researchers' coming to the town was an honour for the participants and made them take the session very serious.	All participants were smallholders or involved in producer association. They corrected each other when somebody played unlike the role in the real world.	Participants knew each other beforehand but did not have business relations with each other. One observation of real-world leader being dominant in the session.	All observations from and actions by smallholders. It is their perspective of the supply network.

7. Discussion and conclusions

Chapters 5 and 6 presented two studies using gaming simulation to generate and test hypotheses about the organisation of transactions. For both the Trust and Tracing Game and the Mango Chain Game conclusions were drawn based upon the empirical findings. In this chapter the outcomes and experiences of the two studies are compared and discussed. Section 7.1 discusses the experiences with gaming simulation to generate and test hypotheses. Attention is paid to validity and reliability. The section ends with the implication for gaming simulation methodology. Section 7.2 compares the theoretical conclusions of the Trust and Tracing Game with those of the Mango Chain Game and relates them to the main research question presented in Chapter 1: '*What is the influence of social structure on the organisation of transactions in supply networks?*' Section 7.3 presents ideas for future research. Next, Section 7.4 presents implications of the research for the domain of supply chains and networks. Section 7.5 provides implications for other domains. The chapter ends with some concluding remarks in Section 7.6.

7.1 Methodological conclusions and implications

Both studies conducted with the Trust and Tracing Game and Mango Chain Game, respectively, followed the same research method but the process of conducting the research was different. This section discusses the differences between the two processes. The conclusions about the research method lead to implications for the research methodology.

Table 7.1 compares the TTG and MCG on key aspects of the research process. The table illustrates that research with the TTG took more time than that with the MCG. Even the design cycle took more time, while the TTG had initially been developed for learning purposes. The reasons for this difference can be found in the functions used and the number of variables. The TTG started with a broad scope, where the important variables had yet to be found in the design cycle. When the variables were identified after the design cycle (see Section 5.1) there appeared to be over 40 variables to collect, measure and calculate in the empirical cycle, coming from all four levels of the Williamson framework (Figure 4.2). As with any research, the more variables to be tested, the more data are required to get significant results. In contrast to this, the MCG used an analytical model with 9 variables based on theory. The number of variables measured was slightly higher, but they could be mapped, most of them even in advance, to the 9 variables. Therefore, fewer sessions were needed.

7.1.1 Validity

Chapter 5 and 6 both end with a table and short discussion of the validity of the gaming simulation tool. The discussions are based upon the four criteria of validity according to Raser (1969): psychological reality, structural validity, process validity and predictive validity. The tables (Table 5.15 and Table 6.9) describe the relationship between the six inputs of a session and the four criteria of validity.

Table 7.1. Methodological comparison TTG and MCG.

Aspect	Trust and Tracing Game	Mango Chain Game
Throughput time for design cycle	1 year for test sessions	6 months for building the MCG
Throughput time for empirical cycle	3 years, 1 researcher	5 months, 2 researchers, 1 student
Sessions needed for significant quantitative outcomes	27	5
Number of variables to incorporate	> 40	9
Physical appearance of gaming simulation	Participants sitting behind separated tables and face-to-face meetings in one room	Board game with face-to-face meetings in one room
Typical time needed for a session	15 minutes preparation, 30 - 45 minutes play, 30 - 60 minutes debriefing.	30 minutes preparation, 1,5 - 2 hours of play, 60 minutes debriefing.
Type of hypothesis	Induced and theory-based	Model based on theory
Functions used	Hypothesis generation and testing	Hypothesis testing
Multi-agent simulation	Yes, validated on system-level	No
Selection of participants	For empirical sessions: student groups	All members of a producer association
Real-world implications of participation	None, students were promised in advance that performance in the session would not influence their grade	Participants did not trade with each other in real life, but were members of one producer association
Game leaders	1	1 + 2 assistants + 1 observer.
Location	Class rooms / conference rooms	Community building or shed
Levels tested based on Williamson framework (Figure 4-2)	1, 2, 3 and 4	3 and 4

The ways in which the TTG and MCG model the roles create differences with respect to validity. The roles in the TTG are abstract representations and meant to get the trade process going and to create involvement. The roles in the MCG are based upon real-world roles in the supply network. The situation of the TTG meant that the participants (especially in the empirical cycle) were mostly students without real-world trade experience. This was negative for the psychological reality and predictive validity, but provided a way to test for differences in the trade process with multiple groups who were comparable in knowledge and had differing levels of trust and pre-existing relationships. Table 5.15 shows that all six inputs in the session were tuned in such a way that the process validity prevailed. The predictive validity of the abstract model of a supply network in the TTG, played with students, suffered as a result. Psychological reality was important and could be achieved by creating roles, rules and objectives that created involvement, instead of a reference to a real-world situation. For the MCG the situation was closer to the participants' real world as smallholders played the roles in the simulated supply network. They had knowledge about the producer associations, independent exporters and multinationals. The (Western) retailer was less well known. The close resemblance of roles to the real world and the use of participants from the real supply network increased the validity, not only on the process criterion but also on the predictive and structural criteria. The round-based play just structured the action to make it observable, but interrupted the trade process. The game board provided a means to visualise the supply network that does not exist in the real world.

In summary, the process validity of both gaming simulations was the most important aspect, and psychological reality was required to get the process going. Both the TTG and MCG met these criteria in different ways. The MCG scored more positively on both the structural and predictive validity because of the closer resemblance of the supply network modeled to the real world and the use of real smallholders versus students.

In addition to the influence of the six inputs of a session on the criteria of validity, it is also important to consider what has not been modeled or measured. The following list discusses some issues.

1. Time frame
   Sessions with both the TTG and MCG lasted somewhere between 25 and 90 minutes. Within this time frame the participants had to do business with each other. The short time frame may have emphasised start-up effects in the TTG because opportunistic traders could take their chances. Some of the participants may have invested in the relationship with other participants rather than focusing on immediate business. A short session may not do justice to all the different ways of building business relations.

2. Single session
   Participants were limited to participation in a single session. Repeated sessions have not been conducted. As with the differences in outcomes in game theory (Axelrod, 1984) between single games and tournaments, it could be hypothesised that the advantages of good business relations and the network mode of organisation become more apparent in the long term. Neither the TTG nor the MCG has been played with a change in incentive structures during a session or between repeated sessions with the same participants.

3. Level of analysis
   Both the TTG and MCG measured the success of individual companies / actors in a supply network. Neither gaming simulation assessed the relative success of a particular combination of buyers and suppliers in a supply network. It might be that the network mechanism is necessary to make one complete chain or even the entire network perform better, giving the consumer a better product while charging a good price. The total revenues

of the network could increase, thus raising the total share for all the traders in the network, while maintaining the relative shares.

4. Limited setting of firm processes

There was no production planning, transportation, stock-keeping or other firm process in either gaming simulation. One of the major benefits of close ties is the sharing of information. This information helps to reduce the costs of, for instance, production and stocks. It could be that the traders who set up close ties with trusted trade partners would have been the most profitable ones in a gaming simulation that modelled processes that depend on information. The reader will find more about this idea under 'future research' (Section 7.3).

5. Variance in trading capabilities

While real smallholders participated in the Mango Chain Game, the dominant group of participants in the Trust and Tracing Game were students. It could be the case that students with more trading experience could exploit the trust and embeddedness of less experienced students. Test rounds or other ways of getting inexperienced participants on the learning curve could remove this starting difference in trading capabilities.

6. No trade history

The participants in a session of both gaming simulations did not have prior business relations with each other. The smallholders in Costa Rica were all members of the same producer co-operation, but this is different from having a trade relationship. In the real world supply networks are less likely to emerge in a setting where there are no prior trade relations, though on a dyadic level companies form new business relations all the time. The conclusions should therefore be seen in the light of business relations to be newly established. The participants in the MCG did, however, have real-world trade experience, from which they incorporated the institutional arrangements with them into the sessions. The TTG was such an abstract model of a supply network that the institutions could come from the rules and roles. The governance level emerged from doing business.

While the second reason is a measurement issue, the others are about the validity of the simulation. As with any simulation, a computer or a gaming simulation always models an aspect of the reference system (the real world). Within the limitations of the Trust and Tracing Game and the Mango Chain Game the behaviour of the participants was real, as they were highly involved. The conclusions of Chapters 5 and 6 and the following Section 7.2 may well be criticised because there is no real-world business where information from close ties does not have an influence on firm internal processes, nor are real-world supply networks formed that have no history. The experiments, however, were valid as regards the process aspect and held psychological reality.

The multi-agent simulation of the TTG was validated on the macro-level, which means that the tendencies of runs as a whole are valid, but the individual actions of the agents in the simulation are not necessarily valid representations of individual traders in a TTG session. There were also some differences in assumptions between the multi-agent model and the human game. The most important one was that network shortcuts were made impossible in the multi-agent simulation. This was of no importance for the hypotheses that were tested, but should be taken into account so as to be able to really compare data from the human sessions with those from the computer sessions.

7.1.2 Reliability of gaming simulation as a research method

In addition to validity, the reliability of gaming simulation is of importance. The gaming simulation method makes the researcher develop a new research tool (game) for each specific situation. Every new research tool needs to be calibrated, regardless of whether it is about chemistry or economics. The reliability of the tool needs to be tested in different situations. In the current book the two gaming simulations have been tested.

The qualitative analysis of the TTG showed that similar results emerged when playing a session with similar groups. The question arises as to what the influence is of the designer and game leader(s) on the outcomes of a session. This question is hard to answer based upon the experiences in this research. The PURDUE groups with American MBA participants and the TOMATO groups with similar participants but with two different game leaders illustrate the stability of the TTG to generate similar outcomes using similar inputs. For the MCG, the congruence of the results that made it possible to test hypotheses with just five sessions illustrates its reliability. The MCG yielded a constant quantity of high quality data that were comparable. Using the same game leaders, issues of different game leaders leading to different outcomes have been avoided.

During the design cycle experiments were carried out in other sessions with the so-called 'deadly envelope' introduced in the last round, to create a shock effect as regards the necessity of good information. The analysis of sessions with a 'deadly envelope' showed that the introduction of a new element in the last round of the TTG confused participants. Some participants just ignored the new element and some were so occupied with the change that their trading was interrupted. This led to the conclusion that the deadly envelope was a bad idea for the TTG in the empirical cycle, as the expression of the trade process was troubled by this new element. It would require a separation in data collection between envelopes bought before the announcement of the deadly envelope and after. That posed challenges for the data collection and thus required changes to the facilitation, and more sessions would have been required to compensate for yet another variable. However, in more qualitatively oriented sessions and sessions for training, the deadly envelope may be a valuable addition.

7.1.3 Reliability of data

Doing quantitative analyses on data that are gathered in a gaming simulation presents challenges that are inherent to the method. The experiences from the TTG led to improvements in the design of the MCG. This subsection explains the challenges and improvements.

A complicating factor for data gathering and data analysis is the source of the data. Both in the TTG and MCG the actions and materials in the session are recorded, either by collecting the game materials like money and products and logging their final location or by collecting the contracts. These data are the by-product of the process in the session, and handled by the participants directly. In a lively session participants are highly involved in their role, and some forget about the administration or skip the parts they think they will remember. Therefore a lot of data are incomplete, not correctly filled in or plainly abused.

Based upon the experiences with the qualitative sessions the collection of materials in the TTG was changed to facilitate data collection. Participants were asked to put the envelopes they owned at the end of the session in envelopes together with the post-session questionnaire that contained their name and player ID. Thus, goods could be followed throughout the whole supply network and coupled to the data from questionnaires without changing the session itself. Reconstructing the data and converting it into an analyzable format (i.e. SPSS / Excel) was a disproportionly time-consuming task in the case of the TTG. One of the key problems was that transactions were logged as properties of a product. Thus, the route of a product through the network could be reconstructed. However, the research questions and the data collected with the questionnaires were on the level of the behaviour of the participants as traders in the network. The researchers had to build a custom data analysis tool to aggregate the route of the products into a list of transactions, cheats, trade partners, etc, per agent.

The MCG design was tailored from the start to facilitate data collection for quantitative data analysis. Contracts and transactions were logged as properties of the traders and pre-structured contract forms (see Appendix D and E) contained the variables of interest in a multiple choice or single value type. The number of variables to be recorded during the session was about half of that in the TTG. The structure of the forms and the round-based play meant that the game leaders had more control over people who did not fill in their forms correctly. The questionnaire was separated from the MCG session, giving the interviewers more time to check for missing items in the questionnaires as well. All of these improvements meant that statistical analyses could be conducted on the data from only 5 sessions with 82 transactions in total.

7.1.4 Methodological implications

The experiences with the TTG and MCG led to the conclusion that gaming simulation is a suitable research method for (quantitatively) testing hypotheses in supply networks. It can be used for the broad, inquiring type of research done with the Trust and Tracing Game with a hypothesis generating and hypothesis testing phase, but also for theory-based research, as in the Mango Chain Game. The previous section on reliability and validity showed that the constructs that were claimed to be modelled in the simulations were present. Care should be taken regarding claims about what is modelled.

The multi-agent simulation has been developed to perform sensitivity analyses of variable settings (loads). This project has not reached the point where variable settings selected with the MAS have been tested in a human session. The multi-agent simulation used another structure of variables, due to the point at which the multi-agent simulation design started. At the point of design the last sessions in the design cycle were just finished and the experimental session setup had not been completely defined. Future research should make clear whether MAS really helps to increase the efficiency by reducing the number of sessions needed through selection of interesting loads. This project has proven, however, that it is possible to develop and validate a multi-agent model of a gaming simulation.

The different types of data that the TTG and MCG can generate are hard to obtain using other methods. As Chapter 2 pointed out, gaming simulation is special in that the participants are exposed to a laboratory-like situation that isolates them from the real world (trading) environment. In this laboratory environment, the attention of the participants can be focused on a particular problem, while retaining the full richness of human behaviour. The TTG provided the possibility to study 27 groups with varying levels of social embeddedness in the same environment which is unique compared to any other method. Case studies in 27 comparable supply networks would be hard to find, let alone the time required for doing 27 case studies gathering transaction data, whereas questionnaire research will not provide the detailed transaction-level data. The MCG forced smallholders to make contractual arrangements and transactions in a 'not-for-real' trade situation. Experimenting with making deals in the real-world could be risky for the relationships with their clients. A study of bargaining power and revenue distribution will be very difficult when real-world contracts and transactions have to be captured without interfering too much in the social relations.

Based upon the experiences presented in this book, gaming simulation can be positioned as a research method that facilitates a whole range of data collections. It is possible to acquire data before, during and after a session, enabling it to be coupled with questionnaires and interviews and actual observation of actions. It can analyse differences between participants in one session, testing for differences in backgrounds, or between session, testing for the effects of varying the load and situation in a session, or even the rules, roles, objectives and constraints of the gaming simulation itself. The experiences with the TTG show that this gaming simulation allows the testing of two different hypotheses. The combination of qualitative and quantitative analyses that is possible using gaming simulations makes the method a good candidate for research that requires both. In the methodology used in the MCG and TTG the first cycle (design cycle) was based upon a qualitative approach, while the hypotheses were tested in a quantitative empirical cycle.

7.2 Theoretical conclusions

The TTG operationalised an abstract supply network of a good with a hidden quality attribute. The TTG has been developed as a learning tool originally. Hypotheses were generated during the last series of test sessions in the design cycle. From observations of 15 test sessions, meant to identify the learning effect, conclusions about the participants' behaviour in the sessions were drawn. The results showed that participants used the two modes of organisation, both the network and the market mechanism.

The conclusions of the design cycle were used as induced hypotheses for the empirical cycle. The empirical cycle tested two hypotheses with quantitative data gathered using gaming simulation and pre- and post-session questionnaires. The first hypothesis was induced from the previous observations (*The dominant mode of organisation in the TTG is network, not market*) and could be confirmed for consumers and rejected for traders. Further analysis showed the influence of pre-existing social relations on the course of the action in the sessions. The second hypothesis that was drawn from theory (*High trust between traders in a network reduces transaction costs*) has been rejected. No relationship could be found between the level of trust and the measurable transaction costs from tracing. Further analysis revealed that participants are able to suspect cheats based on other factors than tracing. The findings of the empirical cycle are in line with those from the design cycle, insofar as social structure influenced the organisation of transactions. From the hypothesis generation and based on the literature, however, it was hypothesised that embeddedness and existing social relations (providing trust) would be positive for the economic performance of traders in a session, and that it would reduce transaction costs. In the setting of the TTG this was not the case. For the consumers for whom only points were important at the end of a session, the ones who earned most points did not know the other participants in advance and trusted them less. Well-performing consumers used the network mode of organisation with business partners they did not know in advance. For the traders it was the other way around. Well-performing traders used the market mode of organisation, and tried to form monopolies on goods by cooperating horizontally. In addition to having as many clients as possible it turned out to be gaining a positive result for them to exploit clients that they knew in advance, exploiting trust from pre-existing relations.

The MCG was developed to study bargaining power and revenue distribution in the mango network from Costa Rica to Western consumers with mango producing smallholders as participants. An analytical model (Figure 6.2) was used as a hypothesised model based upon theory.

The design phase resulted in an appropriate gaming simulation. The observations from the design cycle were used to configure the load and situation in the experimental session setup. No induced hypotheses were formulated, as the MCG was part of a larger project that used research questions based on theory (Zuniga-Arias, 2007).

The empirical cycle led to significant results in five sessions. As expected, a lower bargaining power of the buyer (seller) resulted in higher revenue for the seller (buyer). In general, stakeholders with more bargaining power were able to take advantage of the other agents. Higher risk-aversion of the buyers and/or the sellers led to more revenues for the other agents involved in the exchange relationship. In the same vein, long-term contracts in the buyer-seller relationship led to lower revenues (but also reduced risk) for sellers. The latter result was surprising, since contract choice appeared only to be significant for the seller's and not for the buyer's revenue equation. Mango producers turned out to be well aware of the fact that the type of markets in which they operate is mainly based on short-terms contracts. Not working with long-term contracts gave them the opportunity to remain flexible towards changes in demand and supply that they cannot control. Producers were trying to establish long-term relationships, but they could as well rely on repeated short-term contracts with the same partner. The latter type of contracts tends to rely on trust or friendship, thus the network mode of organisation is at play here. An incentive to take into account in the transaction between buyers and sellers is that the latter prefer to rely on short-term contracts that include not only price, delivery time and quality attributes but also stipulate risk-sharing to enable them to take some risks to the detriment of the buyers' revenues. Finally, real-world wealth appeared to have a significant impact on bargaining power.

The conclusions of the MCG are in line with the TTG conclusions insofar that the variables of social structure used in this book (trust, embeddedness, norms and values) clearly shape the organisation of the transactions. In the case of the MCG the focus was on the individual-level characteristics of the smallholders in the network that led to a bargaining power position which influenced the revenue distribution. The tendency to make repeated short-term contracts with the same partner, while transferring the risk to the buyer was negative for the revenues of smallholders. This combined with the trust that comes with the repeated trade partner shows a second case of the network mechanism at work that is not beneficial for the performance indicator for 'money'. Again it seems that the network mechanism is essential in supply networks with independent traders but not in a way that directly leads to more revenues.

The conclusions from the two gaming simulations are at odds with the leading paradigm in literature on supply networks that trust and relations are important for successful business. The conclusions are more in line with neo-classical economic theory, where companies use transactions as the rational result of considering price and product. Are traditional economists right and is the network mode of organisation negative for the performance of companies? The conclusions were consistent for the two different gaming simulations, for the two different populations (students versus real-world smallholders) and for the two types of research questions (generated versus theory-based). On the other hand, the previous section discussed that both gaming simulations left out some variables, and offered a safe laboratory-like environment that limits the width of this conclusion.

Interestingly, the TTG showed that in a setting with two different performance indicators (money for traders and points for consumers), the only positive effect of pre-existing relations was the one for traders who could exploit trusting clients. As has been said in the introduction (Section 1.1) of this book, studies on supply chains and networks show mixed effects on business performance (Van der Vorst, 2004). The next section discusses how and where future research could help to clarify the root causes of these mixed effects, building upon the methodology and experiments of the current book.

Based upon the research as presented in this book, the answer to the research question as formulated in Chapter 1: 'What is the influence of social structure on the organisation of transactions in supply networks?' is that a social structure with trust and embeddedness makes supply networks use the network mode of organisation.

7.3 Future research

The explorative nature of this book means that more questions can be raised for future research. The method of gaming simulation for quantitative research opens new pathways to explore social structure in business environments. This section uses the structure of the analytical model in Figure 5.3 is as the structure for discussion.

First of all the current set-up of independent, dependent and mediating variables is not yet fully explored. The first and most obvious future project would be to do some up scaling with the Trust and Tracing Game to get the number of participants required to get significant conclusions about the role of culture. The Mango Chain Game could be repeated with other nodes in the supply network. All information in the current research has been gathered with smallholders in Costa Rica as participants.

Furthermore new independent variables to so far have been ignored could be introduced. Future sessions of the TTG should incorporate personality, and may try to formalise the pre-existing relationships between participants. Running sessions all over the world should lead to interesting data about how culture and personality moderate the influence of pre-existing social relations in the TTG setting. Repeating the MCG with mango producers from other parts of the world like Ghana could show differences in networks of similar products. The effect of cultural differences could be measured in such games.

Future research could chance what is being modelled. In the current Trust and Tracing Game no company processes like storage, transportation or production are present. In Section 7.1.2 it has been discussed that the influence of information on costs associated with these processes may give well-embedded in these supply networks an advantage. The extension of the TTG with other company processes could give a better insight in the contextual value of trust and embeddedness. In the analytical framework of Figure 5.3 this means new (company-internal) processes on the level 4.

Methodologically future research could focus both on gaming simulation itself as on the combination with other research methods. To start with the last, the integration of gaming simulation with other research methods remains a question to be investigated. Integration of gaming simulation with case studies could help to overcome the generalisability issue. Is it possible to develop a gaming simulation in such a way that empirical data gathered with, for instance, surveys or interviews in a particular network can be brought into a gaming simulation via the load? This way supply networks could be made comparable and the influence of context on the configuration of the network could be tested.

More research is needed to further develop methodology of integrating gaming simulation and multi-agent simulations in a joint methodology. This book has only been a first attempt.

For the method of gaming simulation itself there are questions about techniques. In this book gaming simulations used only paper-and-pencil-type of materials. The popular wave of 'serious gaming' focuses on computer games. Hybrid gaming simulations, combining face-to-face action and computer-supported models are popular especially in training. It would be interesting to see what the influence could be if the TTG is computer-supported. In the literature on multi agent simulations interactions between human decision makers and multi-agent simulations become increasingly popular. The ability to let certain roles be played by agents in for instance a TTG setting deserves further attention. It could help to identify what aspects of negotiation with a human trader are crucial for building business relationships.

Lastly, the research in this book raises questions about the relationship between embeddedness and the network mode of organisation. In the current book embeddedness is used as a measurement for the number and depth of connections between independent firms. Since in this book the firm equals a person, there was no difference in personal and business relationships within the sessions. The concept of embeddedness could also apply to social networks of people who know each other and may be friends. 'Social embeddedness' and 'firm embeddedness' are two different concepts, but analytically often not kept apart. A related issue is trust. Do you have trust in your business partner or in person X who works for company Y? And what type of trust, as different types of trust exist (Hofstede, 2003; Nooteboom, 2002). The network mode of organisation works through trust and reputation. But how exactly does this happen, through what type of trust, and how does reputation relate to the embeddedness of a firm or of a person in a social network? Powell (1990), Williamson(1996, 2000) and Menard and Shirley (2005) have made the network mechanism an accepted concept among academics. Nearly every publication on the network mechanism mentions aspects of social structure like trust, relations, norms and values. But what aspects in the bucket category of social structure apply to the network mechanism only, and which are generic for the three modes of organisation? The questions raised on the relationship between embeddedness and the network mode of organisation can be approached using gaming simulation. The differences between 'social embeddedness' and 'firm embeddedness' can be analysed using hypothesis generation function. Hypotheses on the relevance of aspects of social structure for each of the three modes of organisation (market, network and hierarchy) can be tested quantitatively with gaming simulations similar to the ones used in the current book.

7.4 Implications for the food chains and networks domain

This book gave a proof-of-principle of a methodology for quantitative research into aspects of social structure and embeddedness influencing the organisation of transactions in a supply network, and more specifically a food supply network. In the domain of chains and networks the research methodology used in this book can be a valuable addition to the more common range of methodologies listed in the methodology chapter (Chapter 2).

The findings on the isolated effect of trust and embeddedness on (financial) performance add to the discussion between researchers who find positive effects of trust and the ones who find negative effects. The risk of lock-in (Omta and Van Rossum, 1999) is the dark side of social embeddedness. Circumstances under which this dark side is offset by the positive effects of reduced risks, information sharing and optimisation of the supply network should be sought. By using gaming simulation this book has contributed by investigating the effect of social structure and embeddedness on performance.

Looking at the history of research in the domain of chains and networks this has expanded from optimising material flows in the early years, to analyzing the trade structure based on social relations in recent years. In 2004, Prof. Zylbersztajn opened his keynote speech for the 6th international conference on chain and network management in agribusiness and the food industry as follows: *'Network theory has continued to evolve in recent years, but empirical studies are still lagging behind.'* He, and other authors like Menard and Shirley (2005) and Williamson (2000) agree that studying the organisation of transactions is complex when trust, social networks and embeddedness come into play. Environments of different networks or even sub-branches in a network can be different and the 'noise' from a real-world on-going business can be complicating for research. At the same time traditional economic models (Walrasian auctioneer, Homo economicus) are no longer sufficient since the price mechanism plays a limited role when the network mode of organisation is used (Zylbersztajn, 2004). For the domain of chain and network studies this implies a need for empirical research that includes trust, embeddedness and norms and values in the independent variables explaining the organisation of transactions, governance mechanisms in general, and the structure of supply networks as a whole. The application of gaming simulation as (one of the) research method(s) can be of value for gathering data about the real behaviour of real participants in a simplified 'surrogate' environment, determined by the gaming simulation. In this book a laboratory for chain and network studies has been build.

7.5 Implications for other domains

The domain discussed in this book is supply chains and networks of food products, as the gaming simulations focused on a product with a hidden quality attribute (which was a model for) and a real mango network in Costa Rica. The emphasis on food supply networks raises the question as to what the similarity is with other networks. The dilemma of the Trust and

Tracing Game (do you trust your suppliers or do you spend money on tracing / revealing the truth?) is a dilemma that is not unique to food supply networks. For every transaction where the information is asymmetric this dilemma is present. Almost every product that is not fully transparent has asymmetric information associated in trade. Next to product properties like hidden quality, pollution or defects, the production properties and management skills can be attributed to asymmetric information, as the supplier may know that promised delivery schedules will not be met.

The Mango Chain Game was primarily aimed at the position of smallholders in the supply network. Food supply networks typically have smallholders as local producers. This is also the case in some industrial production networks. Uzzi's (1997) fashion industry research contained smallholders as well. In more general terms, networks producing specialty products can contain many small companies. In networks of knowledge products and the IT sector smallholders have their place too.

Gaming simulation has proven useful for studying the organisation of transactions in supply networks when there is a mix of the network and market mechanism. For supply networks that use the market mechanism only, social structure will not be of influence as all transactions are moderated by the price mechanism. However, Claro *et al.* (2004) showed that even in flower auctions the reputation of a grower is important for the price, which may suggest that strict markets do not exist. In general, research with a gaming simulation is less suitable when there is less communication and interaction to observe. The same holds true for supply networks that are part of a hierarchy. When the actual transactions are governed by internal company rules, the principal-agent dilemma (Eisenhardt, 1988) still occurs. Gaming simulation can be used to study the way rules, roles, incentives and constraints shape the company-internal transactions.

7.6 Concluding remarks

This book showed as a proof-of-principle that gaming simulation is an excellent additional research method for controlled analysis of complex social systems. It also showed that the possibility to have a repeatable experiment within a controlled contextual setting gives insight in socio-economic behaviour in a way that can be approached from multiple bodies of theory. Staying within the framework of one body of theory cannot explain the full richness of human behaviour, thus links have to be made. In the current book the theoretical framework has been used from new institutional economics and some first attempts have been made to link with theory on culture, psychology and other theories in the social sciences. Future research could use gaming simulation as the research method of choice for true interdisciplinary research.

References

Agranoff, R. and B.A. Radin, 1991. The comparative case study approach in public administration. Research in Public Administration, 1: 203-231.

Agrell, P., R. Lindroth and A. Norman, 2002. Risk, information and incentives in telecom supply chains. International Journal of Production Economics, 90: 1-16.

Anderson, J.C. and J.A. Narus, 1990. A Model of Distributor Firm and Manufacturer Firm Working Partnerships. Journal of Marketing, 54 (1): 42-58.

Apostel, L., 1960. Towards the formal study of models in the non formal sciences. Synthese, 12: 125-161.

Argyris, N. and J. Liebeskind, 1999. Contractual commitments, bargaining power, and governance inseparability: Incorporating history into transaction cost theory. Academy of Management Review, 24: 49-63.

Axelrod, R., 1984. The Evolution of Cooperation, New York: Basic Books.

Ayala, J., 1999. Instituciones y Economía. Una introducción al neoinstitucionalismo económico. Fondo de Cultura Económica, México.

Bacharach, S. and E. Lawler, 1984. Bargaining: Power, Tactics, and Outcomes. Jossey-Bass, San Francisco, C.A.

Barreteau, O., 2003. The joint use of role-playing games and models regarding negotiation processes: characterization of associations. Journal of Artificial Societies and Social Simulation, 6(2).

Batten, D., 2000. Discovering Artificial Economics: How Agents Learn and Economies Evolve. Westview Press.

Beers, George, A.J.M. Beulens and J.C. van Dalen, 1998. Chain science as an emerging discipline. In: G.W. Ziggers, J.H. Trienekens and P.J.P. Zuurbier (eds.). Proc. 3rd International Conference on Chain Management in Agribusiness and the Food Industry. Wageningen, The Netherlands. pp 295-308.

Bekebrede, G. and I.S. Mayer, 2006. Build your seaport in a game and learn about complex systems. Journal of Design Research, 5(2): 273-298.

Benbasat, I. and R.W. Zmud, 1999. Empirical research in information systems: the practive of relevance. Management Information Systems Quarterly, 23 (1), pp 3-16.

Bogetoft, P. and H. Olesen, 2004. Design of Production Contracts. Lessons from Theory and Agriculture. Copenhagen Business School Press. Denmark.

Bosse, T. and C.M. Jonker, 2005. Human vs. Computer Behaviour in Multi-Issue Negotiation. Proceedings of the 1st International Workshop on Rational, Robust, and Secure Negotiations in Multi-Agent Systems, IEEE Computer Society Press, pp. 11-24.

Brooks, F., 1975. The Mythical Man-Month: Essays on Software Engineering. Addison-Wesley Professional.

Bryman, A. and E. Bell, 2003. Business Research Methods. Oxford University Press.

Burgoon, J.K., G.M. Stoner, J.A. Bonito and N.E. Dunbar, 2003. Trust and Deception in Mediated Communication. Proceedings of the 36th Annual Hawaii International Conference on System Sciences (HICSS'03) - Track1, p.44.1.

Burt, R.S., 1982. Towards a structural theory of action. New York: Academic Press.

Burt, R.S., 1992. Structural Holes: The Social Structure of Competition. Harvard University Press, Cambridge.

References

Burt, R.S., 1997. The Contingent Value of Social Capital. Administrative Science Quarterly, 42(2): 339-365.

Burt, R.S., 2000. The network structure of social capital. In: B.M. Staw and R.I. Sutton (eds.) Research in Organisational Behavior. Amsterdam; London and New York: Elsevier Science JAI, 2000, pp. 345-423.

Buvik, A. and T. Reve, 2002. Inter-firm governance and structural power in industrial relationships: the moderating effect of bargaining power on the contractual safeguarding of specific assets. Scandinavian Journal of Management, 18: 261-284.

Camps, T., P. Diederen, G.J. Hofstede and G.C.J.M. Vos, 2004. The Emerging World of Chains and Networks: Bridging Theory and Practice. The Hague: Reed Business Information.

Castelfranchi, C., 1998. Modelling social action for AI agents. Artificial Intelligence, 103: 57-182.

Castelfranchi, C. and R. Falcone, 2001. Social Trust: A Cognitive Approach. In: C. Castelfranchi and Y.H. Tan (eds.). Trust and Deception in Virtual Societies. Kluwer Academic Publishers, pp. 55-90.

Castelfranchi, C., R. Falcone and F. de Rosis, 2001. Deceiving in GOLEM: How to strategically pilfer help. In: C. Castelfranchi and Y.H. Tan (eds.). Trust and Deception in Virtual Societies. Kluwer Academic Publishers, pp. 91-110.

Chamberlain, N. and J. Kuhn, 1965. Collective Bargaining, 2nd ed. New York: McGraw-Hill.

Chamberlin, E.H., 1948. An experimental imperfect market. In; Journal of political economy, 56: 95-108.

Chatterji, D., 1996. Accessing external sources of technology, Research Technology Management, 39(2): 48-56.

Checkland, P.B. and J. Scholes, 1991. Soft Systems Methodology in Action. Chichester: John Wiley.

Chemla, G., 2005. Hold-up, stakeholders and takeover threats. Journal of Financial Intermediation, 14: 376-397.

Churchill, G.A. Jr, 1999. Marketing research: methodological foundations (7th edition). New York: The Dryden Press.

Claro, D.P., D. Zylberstajn and S.W.F. Omta, 2004. Managing relationships and be successful: a study of the business network and buyer-supplier relationship in the Dutch potted flower and plant industry. Journal of Chain and Network Science, 8(1): 18-33.

CNP Website, 2007. http://www.cnp.go.cr , July 2007.

Coase, R.H., 1937. The Nature of the Firm. Economica, 2 (1): 386-405.

Coleman, J.S., 1988. Social capital in the creation of human capital. American Journal of Sociology, 94: 95-120.

Coleman, J.S., 1990. Foundations of Social Theory. MA: Harvard University Press.

Cook, K., 1977. Exchange and power in networks of international relationships. Sociological Quarterly, 18(1): 62-82.

Cooper, M.C. and L.M. Ellram, 1993. Characteristics of Supply Chain Management and the Implications for Purchasing and Logistics Strategy. The International Journal of Logistics Management, 4(2): 13-24.

Cooper, M.C., D.M. Lambert and J.D. Pagh, 1997. Supply Chain Management: More Than a New Name for Logistics. The International Journal of Logistics Management, 8(1): 1-14.

Creswell, J.W., 2002. Research Design: Qualitative and Quantitative Approaches (2 edn.), London: Sage.

Corsi, T.M., S. Boyson, A. Verbraeck, S,-P. van Houten, C. Han and J.R. MacDonald, 2006. The real-time Global Supply Chain Game: New educational tool for developing supply chain management professionals. Transportation Journal, 45(3): 61-73.

Crookall, D., 1994. A guide to the literature on Simulation/Gaming. In: D. Crookall and K. Arai (eds.) Simulation and gaming across disciplines and cultures. SAGE publications, pp. 151-177.

Davis, D.D. and C.A. Holt, 1993. Experimental economics. New Jersey: Princeton University Press.

De Caluwé, L., 1997. Veranderen moet je leren. Een evaluatiestudie naar de opzet en effecten van een grootschalige cultuurinterventie met behulp van een spelsimulatie. Tilburg, PhD-thesis. (In Dutch).

De Caluwé, L. and J.L.A. Geurts, 1999. The use and effectiveness of gaming/simulation for strategic culture change. In: Saunders, D. and J. Severn, 1999. Simulations and Games for Strategy and Policy Planning. Londen: Kogan Page.

De Caluwé, L., G.J. Hofstede and V. Peters, 2008. Why do games work? In search of the active substance. Kluwer, Deventer, The Netherlands.

De Rosis, F., C. Castelfranchi, V. Carofoglio and G. Grassano, 2003. Can Computers deliberately deceive? Computational Intelligence 19: 215-234.

Diederen, P.J.M. and H.L. Jonkers, 2001. Chain and Network Studies. KLICT Paper 2415, Den Bosch, The Netherlands.

Dobbelaere, S., 2003. Estimation of price-cost margins and union bargaining power for Belgian manufacturing. International Journal of Industrial Organisation, 22 (10): 1381-1398.

Druckman, D., 1994. The educational effectiveness of interactive games. In D. Crookall and K. Arai (eds.) Simulation and gaming across disciplines and cultures. SAGE publications, pp. 178-187.

Duke, R.D., 1974. Gaming: the future's language. Sage, Beverly Hills / London.

Duke, R.D., 1980. A paradigm for game design. Simulation & Games, 2: 364-377.

Duke, R.D. and J.L.A. Geurts, 2004. Policy games for strategic management. Dutch University Press, Amsterdam, The Netherlands.

Dyer, J.H. and H. Singh, 1998. The Relational View: Cooperative Strategy and Sources of Interorganisational Competitive Advantage. The Academy of Management Review, 23(4): 660-679.

Economides, N., 1996. The economics of networks. International Journal of Industrial Organisation, 13: 673-699.

Eisenhardt, K.M., 1989. Agency Theory: An Assessment and Review. The Academy of Management Review, 14(1): 57-74.

Emerson, R., 1962. Power-dependence relationships. American Sociological Review, 27: 31-41.

Fafchamps, M., 2004. Market institutions in sub-Saharan Africa: theory and evidence. The MIT Press, Cambridge, Massachusetts, London, England.

Fossum, J., 1982. Labor Relations Development, Structure, Process. Business Publications: Plano, Texas.

Gasnier, A., 2008. The patenting paradox, a game-based approach to patent management. Eburon, Delft, The Netherlands.

Gibbs, G.I., 1974. The handbook of games and simulation exercises. Routledge.

Gosen, J. and J. Washbush, 2004. A review of scholarship on assessing experiential learning effectiveness. Simulation & Gaming 35(2): 270-293.

Grandison, T. and M. Sloman, 2000. A Survey of Trust in Internet Applications. IEEE Communications Surveys.

Granovetter, M., 1985. Economic action and social structure: the problem of embeddedness. American Journal of Sociology, 91: 481-510.

Hakansson, H., 1992. Evolution processes in industrial networks. In: B. Axelsson and G. Easton (eds.) Industrial Networks: A New View of Reality, Routledge, London, pp 129-143.

Hakansson, H. and D. Ford, 2002. How should companies interact in business networks? Journal of Business Research, 55: 133-139.

Hakansson, H. and J. Johanson, 1993. The network as a governance structure: interfirm cooperation beyond market and hierarchies. In: G. Grabher (ed.) The Embedded Firm: On the Socioeconomics of Industrial Networks, Routledge, London.

Hanemann, W.M., 1991. Willingness to Pay and Willingness to Accept: How Much Can They Differ? The American Economic Review, 81(3): 635-647.

Harland, C., 1999. Supply Network Strategy and Social Capital. In: R.T.A.J. Leenders and S. Gabbay (eds.) Corporate Social Capital. Kluwer Academic Publishers, MA, USA, pp. 409-431.

Heide, J. and G. John, 1992. Do norms matter in marketing relationships? Journal of Marketing, 56: 32-44.

Hendrikse, G., 2003. Governance in Chain and Networks: A Research Agenda. ERIM Report series Research in Management, Rotterdam, The Netherlands.

Hofstede, G.J., 1992. The WISARD project: Towards good practice in Information Systems research. Proc. EC-workshop 'Statistical Methods in agriculture'.

Hofstede, G.J., 2003. Trust and Transparency in supply chains: a contradiction. Actes du 8ème Colloque AIM Grenoble.

Hofstede, G.J., 2004. Globalisation, culture and netchains. In H. Bremmers, O. Omta, J.H. Trienekens and E. Wubben (eds.) Dynamics in Food Chains. Proceedings of the 6th International Conference on Chain Management in Agribusiness and the Food Industry. Wageningen: Wageningen Pers, pp. 427-434.

Hofstede, G. and G.J. Hofstede, 2005. Cultures and Organisations: Software of the Mind. Third Millennium Edition. New York: McGraw-Hill.

Hofstede, G.J., M. Kramer, S.A. Meijer and J.A.J. Wijdemans, 2003. A Chain Game for Distributed Trading and Negotiation. Production Planning & Control 14(2): 111-121.

Hofstede, G.J. and S.A. Meijer, 2008. Collecting empirical data with games. In: I.S. Mayer and H. Mastik (eds.) Organisation and learning through gaming and simulation: Proceedings of ISAGA 2007. Eduron, Delft, Holland, pp. 111-121.

Hofstede, G.J., P. Pedersen and G. Hofstede, 2002. Exploring Culture: Exercises, Stories and Synthetic Cultures. Yarmouth, Maine: Intercultural Press.

Hofstede, G.J., L. Spaans, H. Schepers, J. Trienekens and A. Beulens (eds.), 2004. Hide or confide: the dilemma of transparency. Reed Business Information. ISBN 90 5901 374 3.

Holloway, G., C. Nicholson, C. Delgado, S. Staal and S. Ehui, 2000. Agroindustrialization through institutional innovation. Transaction costs, cooperatives and milk-market development in the east-African highlands. Agricultural Economics (23): 279-288.

Holt, C.A., 1996. Trading in a pit market. Journal of Economic Perspectives 10, 193-203.

Huizinga, J., 1971. Homo Ludens: a study of the play-element in culture. Beacon Press.

Jiménez, L., 2003. Gerente División Mango, Coonaprosal, Guanacaste, Costa Rica. Personal communication.

Jonker, C.M. and J. Treur, 1999. Formal analysis of models for the dynamics of trust based on experiences. In: F.J. Garijo and M. Boman (eds.) Proceedings of MAAMAW'99. LNAI 1647, pp. 221-232.

Jonker, C.M. and J. Treur, 2001. An agent architecture for multi-attribute negotiation. In: B. Nebel (ed.) Proceedings of the 17th Int. Joint Conf. on AI, IJCAI '01, pp. 1195-1201.

Jonker, C.M., S.A. Meijer, D. Tykhonov and D. Verwaart, 2005a. Multi-Agent Model of Trust in a Human Game. In: P. Mathieu, B. Beaufils and O. Brandouy (eds.) Artificial Economics, A Symposium in Agent-based Computational Methods in Finance, Game Theory and their Applications, AE2005, pp. 91-102. Lecture Notes in Economics and Mathematical Systems, 564.

Jonker, C.M., S.A. Meijer, D. Tykhonov and D. Verwaart, 2005b. Modelling and Simulation of Selling and Deceit for the Trust and Tracing Game. In: C. Castelfranchi, S. Barber, S. Sabater and M. Singh (eds.) Proceedings of the Trust in Agent Societies Workshop, pp. 78-90.

Jøsang, C. and S. Presti, 2004. Analysing the Relationship between Risk and Trust. In: C. Jensen, S. Poslad and T. Dimitrakos (eds.) Trust Management, Proceedings of iTrust 2004. LNCS 2995, pp. 135-145.

Kahneman, D., J. Knetsch and R. Thaler, 1990. Experimental Test of the endowment effect and the Coase Theorem. Journal of Political Economy 98(6): 1325-1348.

Kamann, D.J.F., 1998. Modelling networks: a long way to go. Tijdschrift voor Economische en Sociale Geografie, 68(3): 279-297.

Klabbers, J.H.G., 2003a. Simulation and gaming: introduction to the art and science of design. Simulation and Gaming, 34(4): 488-494.

Klabbers, J.H.G., 2003b. Gaming and simulation: Principles of a science of design. Simulation and Gaming, 34(4): 569-591.

Klabbers, J.H.G., 2006. Guest editorial. Artifact assessment vs. theory testing. Simulation & Gaming, 37(2): 148-154.

Klabbers, J.H.G., 2008. The magic circle: principles of gaming & simulation (2nd edition). Sense Publishers, Rotterdam / Taipei.

Kogut, B., 2000. The network as knowledge: generative rules and the emergence of structure. Strategic Management Journal, 21: 405-425.

Kolb, D.A., 1984. Experiential learning: Experience as the source of learning and development. New Jersey: Prentice-Hall.

Kriz, W.C. and J.U. Hense, 2006. Theory-oriented evaluation for the design of and research in gaming and simulation. Simulation Gaming, 37: 268-285.

Kuhn, T.S., 1962. The Structure of Scientific Revolutions. Univ. of Chicago Press, Chicago.

Kuit, M., I.S. Mayer and M. de Jong, 2005. The INFRASTRATEGO game: an evaluation of strategic behaviour and regulatory regimes in a liberalizing electricity market'. Simulation & Gaming, 36(1): 58-74.

Kumar, S. and A. Seth, 1998. The Design of Coordination and Control Mechanisms for Managing Joint Venture-Parent Relationships. Strategic Management Journal, 19: 579-599.

Lazzarini, S.G., F.R. Chaddad and M.L. Cook, 2001. Integrating supply chain management and network analysis: the study of netchains, Journal on Chain and Network Science, 1(1): 7-22.

Leap, T. and D. Grigsby, 1986. A Conceptualization of Bargaining Power. Industrial and Labor Relations Review, 39: 202-213.

Lecraw, D., 1984. Bargaining power ownership, and profitability of transnational corporations in developing countries. Journal of International Business Studies, 27(5): 877-903.

Lee, J., W. Chen and Ch. Kao, 1998. Bargaining power and the trade-off between the ownership and control of international joint ventures in China. Journal of International Management, 4: 353-385.

Mallon, B. and B. Webb, 2006. Applying a phenomenological approach to games analysis: a case study. Simulation and Gaming 37(2): 3-11.

Mayer, I.S., 2008. Gaming for policy analysis: learning about complex multi-actor systems. In: L. de Caluwe, G.J. Hofstede and V. Peters (eds.) Why do games work? Kluwer.

McConnell, C.R. and S.L. Brue, 2001. Economics. McGraw-Hill Companies.

Meijer, S.A. and G.J. Hofstede, 2003a. Simulations and simulation games in the agro and health care sector. KLICT working paper TR-214, Den Bosch, The Netherlands.

Meijer, S.A. and G.J. Hofstede, 2003b. The Trust and Tracing Game. In: J.O. Riis, R. Smeds and A. Nicholson (eds.) Proc. of the 7th International workshop on experiential learning, IFIP WG 5.7 SIG conference, May 2003, Aalborg, Denmark.

Meijer, S.A. and G.J. Hofstede, 2003c. Simulation games for improving the human orientation of production management. In: G. Zulch, S. Stowasserand H.S. Jagdev (eds.) Current trends in production management. European series in industrial management, Aachen: Shaker-Verlag, 6: 58-64.

Meijer, S.A. and D. Verwaart, 2005. Feasibility of Multi-agent Simulation for the Trust-and-Tracing Game. In: M. Ali and F. Esposito (eds.) Innovations in Applied Artificial Intelligence, Proceedings of IEA/AIE 2005. LNAI 3533, pp. 145-154.

Meijer, S.A, G.J. Hofstede, G. Beers and S.W.F. Omta, 2006. Trust and Tracing game: learning about transactions and embeddedness in the trade network. Journal of Production Planning and Control, 17(6): 569-583.

Meijer, S.A, G.J. Hofstede, G. Beers and S.W.F. Omta, 2008. The organisation of transactions: research with the Trust and Tracing Game. Journal on Chain and Network Science, 8(1): 1-20.

Meijer, S.A., G. Zuniga-Arias and S. Sterrenburg, 2005. Learning experiences with the Mango Chain Game. In: Proceedings of the 9th international workshop on experimental learning in industrial management, May 2005, Espoo, Finland.

Ménard, C. (ed.), 2000. Institutions, Contracts and Organisations: Perspectives from New Institutional Economics. Edward Elgar.

Ménard, C., 2004. The Economics of Hybrid Organisations. Journal of Institutional and Theoretical Economics, 160(3): 345-376.

Ménard, C., 2005. A new institutional approach to organisation. In: C. Menard and M. Shirley (eds.) Handbook of New Institutional Economics. Springer, pp. 281-318.

Ménard, C. and M. Shirley (eds.), 2005. Handbook of New Institutional Economics. Springer.

Millet, D., 2003. The producers' organisations in the Costa Rican mango chain, Governance Structure and Performance. MSc. Thesis. Management Studies Group. Development Economics Group. Wageningen University. The Netherlands.

MIT, 2008. The MIT Beer Game – online. http://beergame.mit.edu.

Mizuta, H. and Y. Yamagata, 2005. Gaming simulation of the international CO_2 emission trading under the Kyoto protocol. In: Agent-Based Simulation: From Modeling Methodologies to Real-World Applications; Post-Proceedings of the Third International Workshop on Agent-Based Approaches in Economic and Social Complex Systems 2004, Springer Tokyo.

Mora Montero, J., 2002. Guia para el cultivo del mango (Mangifera indica) en Costa Rica. Ministerio de Agricultura y Ganadería, ISBN 9968-877-01-8.

Mora, J., 2004. Gerente General Programa Nacional de Mango. Personal communication. Ministry of Agriculture, Costa Rica.

Moss, S. and B. Edmonds, 2005. Sociology and Simulation: Statistical and Qualitative Cross-Validation, American Journal of Sociology, 110(4): 1095-1131.

Muthoo, A., 2000. A Non-Technical Introduction to Bargaining Theory. World Development, 1 (2): 145-166.

Muthoo, A., 2001. The economics of bargaining. In: Knowledge for Sustainable Development: An Insight into the Encyclopedia of Life Support Systems. UNESCO and EOLSS: EOLSS Publishers Co. Ltd, 2002.

Muthoo, A., 2002. Bargaining Theory with Applications. Cambridge University Press.

Nash, J., 1950. The bargaining problem. Econometrica, 18: 155-162.

Nooteboom, B., H. Berger and N.G. Noorderhaven, 1997. Effects of trust and governance on relational risk. Academy of Management Journal, 36: 794-829.

Nooteboom, B., 2002. Trust: forms, foundations, functions, failures and figures. Cheltenham: Edward Elgar.

Noy, A; D.R. Raban and G. Ravid, 2006. Testing social theories in computer-mediated communication through gaming and simulation. Simulation & Gaming, 37(2): 174-194.

Omta S.W.F. and W. Van Rossum, 1999. The management of social capital in R&D collaborations. In: Corporate Social Capital and Liability. Kluwer Academic Publishers, Boston.

Omta, S.W.F., J.H. Trienekens and G. Beers, 2001. Chain and network science: a research framework. Journal on Chain and Network Science, 1(1): 1-6.

Peters, V., G. Vissers and G. Heijne, 1998. The validity of games. Simulation & Gaming, 29(1): 20-30.

Powell, W.W., 1990. Neither market nor hierarchy: network forms of organisation. Research in Organisational Behavior, 12: 295-336.

Powell, W.W., 1996. Inter-Organisational Collaboration in the Biotechnology Industry. Journal of Institutional and Theoretical Economics, 120(1): 197-215.

Ramchurn, S.D., D. Hunyh and N.R. Jennings, 2004. Trust in Multi-Agent Systems. Knowledge Engineering Review.

Raser, J.C., 1969. Simulations and society: an exploration of scientific gaming. Allyn & Bacon, Boston.

Reynolds, A., 1998. Confirmatory program evaluation: a method for strengthening causal inference. The American journal of evaluation, 19: 203-221.

Roelofs, A.M.E., 2000. Structuring Policy Issues, Testing a mapping technique with gaming / simulation. PhD thesis Katholieke Universiteit Brabant, The Netherlands.

Rouchier, J. and S. Robin, 2006. Information perception and price dynamics in a continuous double auction. Simulation and Gaming, 37(2): 195-208.

Roumasset, J., 1995. The nature of the agricultural firm. Journal of Economic Behavior & Organisation, 26 (2): 161-177.

Rousseau, D.M., S.B. Sitkin, R.S. Burt and C. Camerer, 1998. Not so different after all: a cross-discipline view of trust. Academy of Management Review, 23(3): 393-404.

Rubinstein, A. and A. Wolinsky, 1985. Equilibrium in a market with sequential bargaining. Econometrica, 53(5): 113-1150.

Rubinstein, A., 1982. Perfect equilibrium in a bargaining model. Econometrica, 50 (1): 97-109.

SEPSA, 2001. Boletín Estadístico, Sector Agrícola Costarricense. Ministerio de Agricultura y Ganadería, San José, Costa Rica.

Simchi-Levi, D., P. Kaminski and E. Simchi-Levi, 2000. Designing and managing the supply chain: concepts, strategies, and case studies. New York: Irwin McGraw-Hill.

Slichter, S., 1940. Impact of Social Security Legislation Upon Mobility and Enterprise. American Economic Review, 30: 44-77.

Sterrenburg, S. and G. Zuniga-Arias, 2004. The Costa Rica Chain Game. In: G.J. Hofstede, S.A. Meijer, J.O. Riis and R. Smeds (eds.) Proceedings of the 8th international workshop on experimental learning in chain and networks, 24-27 May 2004, Wageningen, The Netherlands.

Stinchcombe, A., 1985. Contracts as hierarchical documents. In: A.L. Stinchcombe and C. Heimer (eds.) Organisation theory and project management. Bergen: Norwegian University Press.

Strauss, A.L. and J.M. Corbin, 1998. Basics of qualitative research; techniques and procedures for developing grounded theory (2nd edition), Sage, London.

Stoop, S., 2008. The chess parable, decision making & war illustrating the value of playing games for decision makers. In: L. de Caluwe, G.J. Hofstede and V. Peters (eds.) Why do games work? Kluwer.

Tan, K.C., 2001. A framework of supply chain management literature. European Journal of Purchasing & Supply Management, 7: 39-48.

Teece, D.J., 1998. Capturing value from knowledge assets: the new economy, markets for know-how and intangible assets. California Management Review, 40: 55-79.

Tykhonov, D, C.M. Jonker, S.A. Meijer and D. Verwaart, 2008. Agent-Based Simulation of the Trust and Tracing Game for Supply Chains and Networks. Journal of Artificial Societies and Social Simulation, 11(3): 1.

Uzzi, B., 1997. Social Structure and Competition in Interfirm Networks: The Paradox of Embeddedness. Administrative Science Quarterly, 42: 35-67.

Van Laere, J., G.J. de Vreede and H.G. Sol, 2000. Supporting Intra-Organisational Distributed Co-Ordination at the Amsterdam Police Force. In: HICSS '00: Proceedings of the 33rd Hawaii International Conference on System Sciences-Volume 1.

Van Liere, D.W., L. Hagdorn, M.R. Hoogeweegen and P.H.M. Vervest, 2004. The business networking game – an experimental research tool for analysing modular business network structures. In: G.J. Hofstede, S.A. Meijer, J.O. Riis and R. Smeds (eds.) Proceedings of the 8th international workshop on experimental learning in chain and networks, 24-27 May 2004, Wageningen, The Netherlands.

Van Strien, P.J., 1985. Praktijk als wetenschap. Assen: van Gorcum. (in Dutch).

Verduyn, T.M., 2004. Dynamism in supply networks: actor switching in a turbulent business environment. TRAIL Dissertation series 2004/5, TNO, The Netherlands.

Vorst, J.G.A.J. van der, 2004. Traceability: state of the art in seven countries In: G.J. Hofstede, L. Spaans, H. Schepers, J.H. Trienekens and A.J.M. Beulens (eds.) Hide or Confide? The dilemma of transparency. 's Gravenhage : Reed Business Information, pp. 73-80.

Ward, D. and H. Hexmoor, 2003. Towards Deception in Agents. Proc. of AAMAS '03.

Wenzler, I., 2003. The portfolio of 12 simulations within one company's transformational change initiative. In: K. Arai (ed.) Social Contributions and Responsibilities of Simulation & Gaming. Chiba, Japan.

Williamson, O.E., 1975. Markets and hierarchies, analysis and antitrust implications: a study in the economics of internal organisation. New York, Free Press.

Williamson, O.E., 1979. Transaction Cost Economics: The Governance of Contractual Relations. Journal of Law and Economics, 22: 233-261.

Williamson, O.E., 1985. The Economic Institutions of Capitalism. New York: The Free Press.

Williamson, O.E., 1996. The Mechanisms of Governance. Oxford: Oxford University Press.

Williamson, O.E., 1998. Transaction Cost Economics: How It Works; Where It is Headed. De Economist, 146(1), Springer Netherlands.

Williamson, O.E., 2000. The New Institutional Economics: Taking Stock, Looking Ahead. Journal of Economic Literature, 38(3): 595-613.

Wooldridge, M. and N. Jennings, 1995. Intelligent agents: theory and practice. Knowledge engineering review, 10(2): 115-152.

Yan, A. and B. Gray, 2001. Negotiating control and achieving performance in international joint ventures: a conceptual model. Journal of international Management, 7: 295-315.

Yin, R.K., 2003. Case study research: design and methods. Sage Publications, London, United Kingdom.

Zingales, L., 1998. Corporate Governance. In: J. Eatwell (ed.), The New Palgrave. Macmillan, London, United Kingdom.

Zuniga-Arias, G., 2007. Quality management and strategic alliances in the mango supply chain from Costa Rica : an interdisciplinary approach for analysing coordination, incentives and governance. Wageningen Academic, Wageningen, The Netherlands.

Zuniga-Arias, G., S.A. Meijer, R. Ruben and G.J. Hofstede, 2007. Bargaining power and revenue distribution in the Costa Rican mango supply chain: a gaming simulation approach with local producers. Journal on Chain and Network Sciences 7(2): 143-160.

Zylbersztajn, D., 2004. Organisation of firm networks: six critical points for empirical analysis. Editorial in Journal on Chain and Network Sciences, 4(1): 1-6.

Appendices

Appendix A.
Accounts of WUR sessions with Trust and Tracing Game

The courses 'Supply Chain Management' (SCM) and 'Food Safety Economics' (FSE) have an international audience (60%) and attract students from all types of study programmes. These are production technology or production economics oriented. The game was obligatory for all students.

Session no, Course, number of participants. Typical observations.
1. FSE, 15. Low trust between participants. Professional culture? Dutch are opportunistic and short-cut the chain. All other nationalities delay short-cutting. Many cheats but few traces. Greek producer was not trusted up front, due to his nationality. Two Eastern European consumers followed a low-quality strategy so that they could not be cheated. The few high-quality envelopes they bought were frauds. They felt very bad about it. The Dutch consumers did not care a lot. The others had mixed feelings. More and more non-Dutch people refused to do business with each other anymore due to cheats.
2. FSE/SCM, 9. The chain is kept intact during the first season, although everybody is standing around the producers. The Dutch consumer is the first to walk from the consumer area. All participants are very aware of the cheating possibility. Initial cheating attempts were observed at all stages in the supply chain. Cheating stopped as soon as all traders became aware that their reputation was at stake. The consumers preferred honest suppliers. The Chinese participants have a preference to trade with each other in Chinese language. The others all communicate in English. Trade is not chaotic, thus cheating behaviour could be observed by consumers by watching the trade closely.
3. SCM, 9. This session was characterised by sitting consumers. Nobody moved physically, thus trade was very ordered. The producers (who know each other well) cooperate to force the middleman to pay more. Dutch friends who are producer and consumer shout to each other that they should start an internet market and that they should trade directly, but nobody does it. Traces are discouraged by the retailers. One of them sings 'shame, shame, shame on you!' when a trace discovers she did not cheat. The good overview of consumers on what happens in the chain discourages traders from cheating. Traders expressed the view that they didn't want a bad reputation among their study mates.
4. SCM, 13. The game had 2/3 Dutch participants and a few others (2 Central America, 1 French and 2 German). Quickly the supply chain split up into a Dutch chain and an international chain, formed by the non-Dutch participants. The two Germans had a consumer role and bought from the international chain and occasionally from the Dutch chain if a retailer offered them the goods in a relaxed way. If retailers were pushy or stressed, the Germans said they didn't trust them.
5. SCM, 20. Chaotic game with 10 Dutch, and 10 non-Dutch participants from all over the world. The production was in Dutch hands completely. Producers 1 and 2 concentrate on

serving anybody who comes to their table, whether middleman or consumer. Producer 3 tries to supply exclusively to the 3 retailers. Two non-Dutch retailers supply the non-Dutch consumers who stick to the consumer area. Dutch consumers are found mostly at the producers 1 and 2, as they are friends in real life. Many cheats are found. Producer 3 switches to honesty to ensure his clientele. Producers 1 and 2 keep cheating.

6. SCM, 13. This was the game with 3 French girls in the consumer role and 1 as middleman, mentioned in the introduction of this paper. The rest of the supply chain was in Dutch hands. Retailer 2 devoted herself to serve the French consumers, being supplied by Producer 1 who is a friend of hers. She dares to guarantee that she is truthful as she trusts P1 to give her the right quality. All others trade chaotically. The Dutch soon switch to speaking Dutch.

Appendix B.
List of sessions with the Trust and Tracing Game

Number	Year, place and number	Sample	#Producers	#Middlemen	#Retailers	#Consumers	#Participants	#Envelopes Sold By P	#Envelopes Bought By M	#Envelopes Bought By R	#Envelopes Bought By C
1	2005 WUR SCM course, session 1	1									
2	2005 WUR SCM course, session 2	1									
3	2005 WUR SCM course, session 3	1									
4	2005 WUR SCM course, session 4	1									
5	2005 WUR SCM course, session 5	1									
6	2005 RU, SCM course	1									
7	2005 WUR Food Safety Economics course	1									
8	2006 WUR Food Safety Economics course	2	4	4	4	9	21	148	33	32	150
9	2006 KSV knowledge session	3	4	4	4	8	20	120	46	31	130
10	2006 Larensteijn SCM week	3	2	1	2	4	9	82	6	21	116
11	2005 INHOLLAND SCM course, session 1	2	3	3	3	14	23	134	46	47	170
12	2005 INHOLLAND SCM course, session 2	2	3	3	3	9	18	105	79	69	58
13	2005 INHOLLAND SCM course, session 3	3	2	1	2	3	8	2	2	5	70
14	2005 INHOLLAND SCM course, session 4	2	3	2	2	6	13	0	0	0	0
15	2005 INHOLLAND SCM course, session 5	2	3	3	3	9	17	134	85	61	146
16	2005 INHOLLAND SCM course, session 6	2	2	2	2	5	11	79	9	37	70
17	2005 INHOLLAND SCM course, session 7	2	2	2	2	5	11	48	48	38	57
18	2005 INHOLLAND SCM course, session 8	2	2	2	2	5	11	85	32	59	83
19	2005 INHOLLAND SCM course, session 9	2	3	2	3	6	14	91	4	21	93
20	2006 Imagineering teachers	2	3	2	2	6	13	127	11	23	138
21	2006 RU, SCM course	1+2	3	3	3	7	16	143	36	12	118
22	2006 WUR SCM course, session 1	2	2	2	2	4	10	40	43	36	24
23	2006 WUR SCM course, session 2	2	3	2	2	5	12	138	43	18	124
24	2006 WUR SCM course, session 4	2	4	4	4	8	20	113	98	77	108
25	2006 INHOLLAND SCM course, session 1	2	2	1	2	4	9	38	17	21	43
26	2006 WSM, Purdue university MBA Group	2	2	2	1	3	8	23	19	9	25
27	2005 KUB SCM course	2	3	3	3	8	17	102	47	37	93
28	2005 TOMATO session Venlo venue 1 Sebas	1+3	3	3	3	6	15	141	62	46	134
29	2005 TOMATO session Venlo venue 2 Gert Jan	1									

Number	Year, place and number	Sample	#Producers	#Middlemen	#Retailers	#Consumers	#Participants	#Envelopes Sold By P	#Envelopes Bought By M	#Envelopes Bought By R	#Envelopes Bought By C
30	2005 Agrotechnology welcome session 7pp	3	2	1	2	3	8	69	0	69	56
31	2005 WSM, Purdue university MBA Group	1+3	3	3	3	12	21	115	23	24	112
32	2003 WSM, Purdue university MBA Group	1									
33	2005 Scholierenconferentie session 1	1									
34	2005 Scholierenconferentie session 2	1									
35	2006 INHOLLAND SCM course, session 2	2	3	3	3	6	15	145	137	123	127
36	2006 INHOLLAND SCM course, session 3	4	3	3	3	8	17	118	31	17	117
37	2006 INHOLLAND SCM course, session 4	4	3	2	2	6	13	122	16	20	118
38	2006 INHOLLAND SCM course, session 5	4	3	2	2	5	12	111	29	13	110
39	2006 Larensteijn SCM course	4	3	3	3	8	17	86	29	22	89
40	2006 INHOLLAND SCM course, session 6	4	2	2	2	5	11	74	12	0	62
41	2006 INHOLLAND SCM course, session 7	4	3	3	3	11	20	100	21	16	120
42	2005 Kids session with primary school children	1									
43	2003 Ministry of Agriculture knowledge centre	1									

The organisation of transactions

Appendix C.
TTG Questions asked in the questionnaires

Pre questionnaire:

Your name:

Your gender: Female   Male

Your year of birth: 19 ..

Your country of birth:

Your home country (if different):

Your study:

How well do you know the other participants? On average I know them:

very well 1...2...3...4...5   not at all

I know some of them much equally well 1...2...3...4...5   I know them better than others.

How much do you trust the other participants?

Most of them can be trusted 1...2...3...4...5   I trust none of them

Post questionnaire:

Did you cheat?

never 1...2...3...4...5   all the time

Did you get away with cheating?

never 1...2...3...4...5   all the time

Did you cooperate with somebody else?

never 1...2...3...4...5   all the time

If so, with whom?

If so, why?

Appendix D.
MCG Questionnaire (translated from Spanish)

Simulation systematisation

Date: _____ Place: _____

1) Role in the simulation: 1. Producer Assoc. 2. Multinational 3. Independent exporter 4. Retailer

2) The simulation is: 1. Very easy to understand and play 2. Easy to understand and play
3. Neither easy nor hard 4. Hard to understand and play 5. Very hard to
understand and play

3) The simulation is: 1. Really close to reality 2. Neither similar nor different to reality 3. It is
not similar to reality

4) How much profit did you make? _____   5) What were your losses? _____

6) Whom did you do business with? _____

7) Why? 1. They were closer 2. I have known them for a very long time 3. I trust them 4. They offer
the best options 5. I didn't have many options 6. They radiate trust

8) During the simulation, your buying and selling expectations. Never changed, always changed. Select
a number between 1 and 10 where 1 is never changed and 10 always changed

Maintain expectations   1 2 3 4 5 6 7 8 9 10

9) Why do you think this happened? _____

10) On a scale of 1 to 10 where 1 is low bargaining power and 10 is high bargaining power. How do
you consider your bargaining power?   1 2 3 4 5 6 7 8 9 10

11) Please mark with an X the number you feel that represents you best on the different items.
In the simulation:

I am very patient	0 1 2 3 4 5 6 7 8 9 10
I am a risk taker	0 1 2 3 4 5 6 7 8 9 10
I have a lot of buyers	0 1 2 3 4 5 6 7 8 9 10
I take time to make important decisions	0 1 2 3 4 5 6 7 8 9 10
I am a wealthy participant in the simulation	0 1 2 3 4 5 6 7 8 9 10
I can negotiate prices	0 1 2 3 4 5 6 7 8 9 10
I am not afraid to break down negotiations	0 1 2 3 4 5 6 7 8 9 10
My main business partner is a good partner	0 1 2 3 4 5 6 7 8 9 10
I keep my word	0 1 2 3 4 5 6 7 8 9 10
I maintain the business expectations	0 1 2 3 4 5 6 7 8 9 10

12) Write down a number between 0 and 10 to describe the perceived bargaining power of the
participants of the simulation

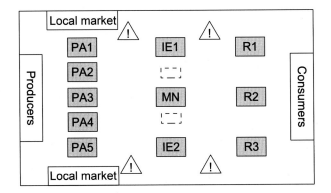

13) Mark with an X to indicate the problems you faced in the simulation?

1 ☐ Little information on which to base decisions
2 ☐ Difficulties to learn the norms and rules of the game
3 ☐ Problems of produce quality in relation to other participants
4 ☐ Few opportunities to do business with other participants
5 ☐ Not knowing anybody beforehand in the simulation
6 ☐ Supply problems
7 ☐ Problems with not meeting the quality requirements of the buyers
8 ☐ Problems when negotiating with other participants
9 ☐ The buyers always try to pay the least
10 ☐ There was a lot of uncertainty in the simulation
11 ☐ I could not look for new partners
12 ☐ I attached myself to a long-term relationship and I got no benefits from it
13 ☐ I always traded in the spot market because I could not sign any contracts

14) Mark with an X the number. In the simulation who do you consider has the most bargaining power

The buyer	0 1 2 3 4 5 6 7 8 9 10
The one with information	0 1 2 3 4 5 6 7 8 9 10
The one with a better economic situation	0 1 2 3 4 5 6 7 8 9 10
The one with many buyers	0 1 2 3 4 5 6 7 8 9 10
The strategic business partner	0 1 2 3 4 5 6 7 8 9 10
The participant with control over activities of other participants	0 1 2 3 4 5 6 7 8 9 10
The business partner with more dependency	0 1 2 3 4 5 6 7 8 9 10
The seller that must sell his perishable product	0 1 2 3 4 5 6 7 8 9 10
The one who contributes most financially	0 1 2 3 4 5 6 7 8 9 10

15) You consider yourself to be 1. Risk taker, 2. Neutral to risk, 3. Risk Adverse

16) Which option do you pick?

        Option A: to get $30, 100% certainty
        Option B: to get $40 with 80% chance, and to get $0 with a chance of 20%

17) How old are you? _____
18) Sex: Masc. Fem.
19) Did you know the people in the simulation beforehand? 1. Yes 2. No
20) Formal Education?_____
21) Main activity: 1. Student, 2. Lecturer, 3. Entrepreneur, 4. Public sector, 5. Mango Producer,
        6. Other: _____
22) Farm size? _____   23) Family size? _____
24) Monthly income? 1. 1,000 to 50,000 colones,   2. 50,001 to 100,000 colones
        3. 100,001 to 250,000 colones, 4. 250,001 to 500,000 colones
        5. More than 500,000 colones

¡MUCHAS GRACIAS!

Appendix E.
MCG Transaction form (translated from Spanish)

Round # _____

Contract

	Red	Yellow
Quantity:	_____	_____ (tons)
Price per ton:	$_____	$_____
Total price:	$_____	$_____

☐ **one round**   ☐ **three rounds**
                     Starts in round:_____

☐ **for ever / buy company**
Note: The buyer has complete control over the company he buys.

Total price: **Form of Payment:**
                ☐ Total price directly paid
$_____   ☐ Each round $_____ for _____ rounds
                ☐ Other form; describe below:

Does the seller keep working for the buyer?
☐ yes   ☐ no   **Salary per round $_____**

Effect of breaking or not   **Who pays the loss of**
fulfilling the contract: **products?:**
☐ nothing   ☐ Buyer
☐ cancel contract and fine $_____   ☐ Seller
☐ fine per round $_____

Agreement for improving quality:_____
_____

Other agreements:_____
_____

Seller: _____   **Buyer:** _____

Appendix F.
Multi-agent simulation results

Table A1. Statistics of simulated games with varying values of trust update. Positive trust update=negative trust update; strategy neutral; initial trust 0.5; initial and minimal honesty 1.0; confidence 0.95.

Trust update	N	H	L	Q	C	G	P	D	H/N	C/H	G/H	(Q+P)/H	P/(H-C)	D/N
0.1	375	250	125	0	29	196	89	0	0.67	0.12	0.78	0.36	0.40	0.00
0.1	366	232	134	0	18	180	84	0	0.63	0.08	0.78	0.36	0.39	0.00
0.1	364	220	144	0	25	167	81	0	0.60	0.11	0.76	0.37	0.42	0.00
Total	1105	702	403	0	72	543	254	0	0.64	0.10	0.77	0.36	0.40	0.00
0.5	376	266	110	0	23	141	62	0	0.71	0.09	0.53	0.23	0.26	0.00
0.5	465	356	109	0	18	186	69	0	0.77	0.05	0.52	0.19	0.20	0.00
0.5	500	369	131	0	21	208	58	0	0.74	0.06	0.56	0.16	0.17	0.00
Total	1341	991	350	0	62	535	189	0	0.74	0.06	0.54	0.19	0.20	0.00
0.9	500	385	115	0	17	201	47	0	0.77	0.04	0.52	0.12	0.13	0.00
0.9	500	386	114	0	26	177	47	0	0.77	0.07	0.46	0.12	0.13	0.00
0.9	500	396	104	0	11	206	50	0	0.79	0.03	0.52	0.13	0.13	0.00
Total	1500	1167	333	0	54	584	144	0	0.78	0.05	0.50	0.12	0.13	0.00

Legend:
N	Number of transactions
H	Number of high quality transactions
L	Number of low quality transactions
Q	Number of certification traces
C	Number of certified transactions
G	Number of guarantees
P	Number of post-transaction traces
D	Number of deceptions
H/N	Proportion of high quality transactions
C/H	Proportion of certified transactions wrt high quality transactions
G/H	Proportion of guarantees wrt high quality transactions
(Q+P)/H	Total tracing frequency wrt high quality transactions
P/(H-C)	Post-transaction tracing frequency
D/N	Deceit frequency

Table A2. Statistics of simulated games with varying values of initial trust and honesty for different negotiation strategies. Minimal honesty=initial honesty; positive trust update=0.3, negative trust update=1.0; confidence=0.95. Tables 5, 6, and 7 present aggregated views of the data.

Initial trust	Honesty	N	H	L	Q	C	G	P	D	H/N	C/H	G/H	(Q+P)/H	P/(H-C)	D/N
Opportunistic															
1.0	1.0	200	164	36	0	0	69	7	0	0.82	0.00	0.42	0.04	0.04	0.00
1.0	0.5	200	147	53	0	0	64	8	20	0.74	0.00	0.44	0.05	0.05	0.10
1.0	0.0	200	124	76	0	0	61	6	39	0.62	0.00	0.49	0.05	0.05	0.20
0.5	1.0	200	133	67	0	10	81	49	0	0.67	0.08	0.61	0.37	0.40	0.00
0.5	0.5	200	121	79	0	1	68	34	13	0.61	0.01	0.56	0.28	0.28	0.07
0.5	0.0	200	131	69	0	3	80	55	31	0.66	0.02	0.61	0.42	0.43	0.16
0.0	1.0	200	116	84	0	19	78	69	0	0.58	0.16	0.67	0.59	0.71	0.00
0.0	0.5	200	107	93	0	8	80	72	10	0.54	0.07	0.75	0.67	0.73	0.05
0.0	0.0	200	99	101	0	4	72	69	14	0.50	0.04	0.73	0.70	0.73	0.07
Quality minded															
1.0	1.0	200	154	46	5	5	32	6	0	0.77	0.03	0.21	0.07	0.04	0.00
1.0	0.5	200	148	52	60	63	22	9	9	0.74	0.43	0.15	0.47	0.11	0.05
1.0	0.0	200	155	45	34	35	54	7	24	0.78	0.23	0.35	0.26	0.06	0.12
0.5	1.0	200	133	67	6	23	82	38	0	0.67	0.17	0.62	0.33	0.35	0.00
0.5	0.5	200	135	65	67	91	25	17	12	0.68	0.67	0.19	0.62	0.39	0.06
0.5	0.0	200	144	56	49	73	54	34	18	0.72	0.51	0.38	0.58	0.48	0.09
0.0	1.0	200	122	78	2	18	99	71	0	0.61	0.15	0.81	0.60	0.68	0.00
0.0	0.5	200	116	84	65	79	34	31	2	0.58	0.68	0.29	0.83	0.84	0.01
0.0	0.0	200	132	68	53	73	52	50	9	0.66	0.55	0.39	0.78	0.85	0.05
Thrifty															
1.0	1.0	200	65	135	0	1	32	4	0	0.33	0.02	0.49	0.06	0.06	0.00
1.0	0.5	200	7	193	4	4	3	1	0	0.04	0.57	0.43	0.71	0.33	0.00
1.0	0.0	200	24	176	24	24	0	0	0	0.12	1.00	0.00	1.00		0.00
0.5	1.0	200	24	176	0	0	17	4	0	0.12	0.00	0.71	0.17	0.17	0.00
0.5	0.5	200	7	193	4	4	2	2	0	0.04	0.57	0.29	0.86	0.67	0.00
0.5	0.0	200	16	184	16	16	0	0	0	0.08	1.00	0.00	1.00		0.00
0.0	1.0	200	17	183	1	3	14	12	0	0.09	0.18	0.82	0.76	0.86	0.00
0.0	0.5	200	19	181	13	13	6	5	0	0.10	0.68	0.32	0.95	0.83	0.00
0.0	0.0	200	37	163	36	37	0	0	0	0.19	1.00	0.00	0.97		0.00

Table A3. Aggregated statistics of 3×9 simulated games with varying values of initial trust. The ratios H/N, ..., D/N are calculated on the rows of the table, so they are a weighted average of the ratios in Table A2, applying the ratio's denominator as weight factor.

Initial trust	N	H	L	Q	C	G	P	D	H/N	C/H	G/H	(Q+P)/H	P/(H-C)	D/N
1.0	1800	988	812	127	132	337	48	92	0.55	0.13	0.34	0.18	0.06	0.05
0.5	1800	844	956	142	221	409	233	74	0.47	0.26	0.48	0.44	0.37	0.04
0.0	1800	765	1035	170	254	435	379	35	0.43	0.33	0.57	0.72	0.74	0.02

Table A4. Aggregated statistics of 3×9 simulated games with varying values of honesty (minimal honesty and initial honesty both set equal to the value in the first column).

Honesty	N	H	L	Q	C	G	P	D	H/N	C/H	G/H	(Q+P)/H	P/(H-C)	D/N
1.0	1800	928	872	14	79	504	260	0	0.52	0.09	0.54	0.30	0.31	0.00
0.5	1800	807	993	213	263	304	179	66	0.45	0.33	0.38	0.49	0.33	0.04
0.0	1800	862	938	212	265	373	221	135	0.48	0.31	0.43	0.50	0.37	0.08

Table A5. Statistics of simulated games with varying values of trust update. Strategy=neutral; initial trust=0.5; initial and minimal honesty=1.0; confidence=0.95.

Delta trust Pos	Neg	N	H	L	Q	C	G	P	D	H/N	C/H	G/H	(Q+P)	P/H	D/N(H-C)
0.05	0.5	500	67	433	21	30	22	9	1	0.13	0.45	0.33	0.45	0.24	0.00
0.05	0.5	500	125	375	70	91	23	13	6	0.25	0.73	0.18	0.66	0.38	0.01
0.05	0.5	500	67	433	28	37	19	14	4	0.13	0.55	0.28	0.63	0.47	0.01
Total		1500	259	1241	119	158	64	36	11	0.17	0.61	0.25	0.60	0.36	0.01
0.5	0.5	500	136	364	59	82	33	15	8	0.27	0.60	0.24	0.54	0.28	0.02
0.5	0.5	500	125	375	24	35	54	24	10	0.25	0.28	0.43	0.38	0.27	0.02
0.5	0.5	500	128	372	28	47	36	14	8	0.26	0.37	0.28	0.33	0.17	0.02
Total		1500	389	1111	111	164	123	53	26	0.26	0.42	0.32	0.42	0.24	0.02
0.1	1.0	500	100	400	17	37	37	22	3	0.20	0.37	0.37	0.39	0.35	0.01
0.1	1.0	500	76	424	24	33	25	14	0	0.15	0.43	0.33	0.50	0.33	0.00
0.1	1.0	500	88	412	30	40	27	17	8	0.18	0.45	0.31	0.53	0.35	0.02
Total		1500	264	1236	71	110	89	53	11	0.18	0.42	0.34	0.47	0.34	0.01
1.0	1.0	500	143	357	37	57	57	24	15	0.29	0.40	0.40	0.43	0.28	0.03
1.0	1.0	500	90	410	27	40	28	10	2	0.18	0.44	0.31	0.41	0.20	0.00
1.0	1.0	500	119	381	25	39	39	20	11	0.24	0.33	0.33	0.38	0.25	0.02
Total		1500	352	1148	89	136	124	54	28	0.23	0.39	0.35	0.41	0.25	0.02

Table A6. Aggregated statistics of 3×9 simulated games with different strategies defined in Section 5.

Strategy	N	H	L	Q	C	G	P	D	H/N	C/H	G/H	(Q+P)/H	P/(H-C)	D/N
Opportunist	1800	1142	658	0	45	653	369	127	0.63	0.04	0.57	0.32	0.34	0.07
Quality-mdd	1800	1239	561	341	460	454	263	74	0.69	0.37	0.37	0.49	0.34	0.04
Thrifty	1800	216	1584	98	102	74	28	0	0.12	0.47	0.34	0.58	0.25	0.00

Table A7. Statistics of simulated games with different mixed strategies. Minimal honesty=initial honesty=0.5; initial trust=0.5; positive trust update=0.3, negative trust update=1.0; confidence=0.95.

N	H	L	Q	C	G	P	D	H/N	C/H	G/H	(Q+P)/H	P/(H-C)	D/N
Opportunistic and quality-minded strategies													
500	195	305	13	26	101	49	40	0.39	0.13	0.52	0.32	0.29	0.08
500	169	331	8	22	72	33	24	0.34	0.13	0.43	0.24	0.22	0.05
500	213	287	6	20	113	58	50	0.43	0.09	0.53	0.30	0.30	0.10
Total 1500	577	923	27	68	286	140	114	0.38	0.12	0.50	0.29	0.28	0.08
Opportunistic and thrifty strategies													
500	64	436	4	8	35	7	6	0.13	0.13	0.55	0.17	0.13	0.01
500	59	441	0	5	32	11	6	0.12	0.08	0.54	0.19	0.20	0.01
500	60	440	2	7	33	12	11	0.12	0.12	0.55	0.23	0.23	0.02
Total 1500	183	1317	6	20	100	30	23	0.12	0.11	0.55	0.20	0.18	0.02

Summary

This book studies the organisation of transactions in supply networks. More specifically it investigates the influence of social structure on the mode of organisation in supply networks. To gain new insights, the results in this book have been gathered using gaming simulation as a research method. As this is a new application of gaming simulation, special attention is paid to the methodological implications.

Food supply chains and networks span a whole series of firms from grower to consumer. Depending on the product traded and the market in which it is consumed the grower and consumer can be located in countries thousands of kilometres apart. However, the food still needs to arrive in a perfect condition at the consumer, passing through several companies in the supply network. Transactions need to be made between the subsequent companies that trade the product. The way in which the transactions are organised can be any mix of three modes of organisation, namely market, network and hierarchy. This book concentrates on market and network. The market mechanism uses the price as a control mechanism. Network uses trust and reputation for this.

Transactions are made by people. People who trade with each other may have known one another for a long time. Their companies may have business ties. And the traders will have certain norms and values about what is appropriate behaviour in trade. The interpersonal relations, business ties and norms and values influence trade behaviour in a supply network. This behaviour will influence the mode of organisation of the network. Lazzarini *et al.* (2001) call this *social structure*, as a bucket category of variables from the social sciences that explain interpersonal and business relations and the norms and values of the supply network as a whole. Trust between traders is the most prominent interpersonal variable. Business relations are expressed in a level of embeddedness as a measure of the density and the strength of ties between businesses. Norms and values in a network can be related to the culture traders come from. *Social structure* covers a broad range of concepts from sociology to social psychology and network theory. This book focuses on the major variables trust and embeddedness, with norms and values as an important context.

Gaming simulation is commonly used as a training or learning tool. This book, however, uses gaming simulation as a research method. The methodological contribution of this book is to use gaming simulation as a lab environment to generate and test hypotheses using both qualitative and quantitative data in the domain of supply chains and networks. Chapter 2 discusses the methodological issues of gaming simulation as a research method. The first section (2.1) describes what gaming simulation is and gives 6 inputs for a session with a gaming simulation. Section 2.2 argues that by using gaming simulation, researchers can study the behaviour of real people playing a role of interest for research in a simulated environment, based upon the characteristics of gaming simulation. Research purposes are less common among gamers. Section 2.3 first describes the non-research purposes, while Section

2.4 discusses the purposes for research. Three types of research purpose are distinguished, namely hypothesis generation, hypothesis testing and sensitivity analysis when combined with multi-agent simulation. Research with gaming simulation is often positioned in the design sciences, which means that the effect of the gaming simulation as a design on changes in the real world is tested. This book positions gaming simulation in the analytical sciences, to study phenomena in the real world.

Section 2.5 positions gaming simulation among other research methods common in the domain of supply chains and networks. The influence of social structure on the organisation of transactions can be studied in a single or small set of supply networks using case studies, to provide in-depth observations of actions and the surrounding context. The generalisability of detailed case studies is a complicated matter. Furthermore, it is hard to observe the actual actions in a case study research. Questionnaire research overcomes the generalisability issue of case studies, though lacks the in-depth knowledge of a subject within its contextual variables. Surveys do not observe actual actions either.

The validity and reliability of gaming simulation is discussed in Section 2.6 and is based on the work of Raser (1969) who identified four criteria for validity: psychological reality, process, structural and predictive validity. Each of the four criteria has been used in this book.

Chapter 3 presents the research method used in this study. It consists of four interconnected cycles. The first is the design cycle in which the gaming simulation is developed and tested. The test sessions provide insight into the structure and important variables of the problem studied. The hypothesis generation function can be done using the design cycle. The outcomes of the design cycle are induced hypotheses (based upon the test sessions) and the gaming simulation. Both are inputs for the empirical cycle in which a structured experimental set-up results in game sessions, which provide the data to be analysed.

The other two cycles are support cycles. The first is the multi-agent design cycle in which a multi-agent version of the gaming simulation is built. The second one is the multi-agent simulation cycle in which experiments can be conducted to verify the multi-agent model or to draw conclusions. Multi-agent simulation could in the future provide ways to select interesting variable settings to play with human participants. The multi-agent simulation is validated against conclusions from the empirical cycle.

Chapter 4 discusses the reference theories on which this book is based. Section 4.1 presents theories on supply chains and networks needed for the domain of study. Section 4.2 discusses new institutional economics used as the main theoretical framework for analysis of the results of the two gaming simulations. Central elements are the four-level framework by Williamson (2000) that links levels of analysis from culture to day-to-day operations, and the modes of organisation, namely network, market and hierarchy. Section 4.3 discusses the fact that there are other explanatory theories used for the two specific gaming simulations. These theories are discussed in the subsequent chapters.

Two custom-built gaming simulations each study an aspect of the influence of social structure on the mode of organisation of transactions. Chapter 5 presents the first one, called the Trust and Tracing Game (TTG). The TTG assessed the influence of trust and embeddedness on the choice between the network and the market mode of organisation. The TTG is a paper-based gaming simulation of a supply network of a product with a hidden quality attribute. Participants face the dilemma of whether to rely on trust or tracing when confronted with a possible cheat. Section 5.1 describes the fact that the TTG was originally designed to be a learning tool by the researchers who started this project. The TTG operationalised an abstract supply network of a good with a hidden quality attribute. Hypotheses were generated during the last series of test sessions in the design cycle. From observations of 15 test sessions, intended to identify the learning effect, conclusions about the participants' behaviour in the sessions were drawn. The results showed that participants used the two modes of organisation, both the network and the market mechanism.

Section 5.2 uses the observations and variables identified in 5.1 as inputs to the empirical cycle for the quantitative analysis of 27 additional sessions as induced hypotheses and list of variables to be collected. The quantitative analysis proved that the mode of organisation in the Trust and Tracing Game was *network* for the financially well-performing traders and *market* for the well-performing consumers. Social structure manifested itself in trust and embeddedness influencing the organisation of transactions. Generally trust and embeddedness were detrimental to the (financial) performance in the setting of the Trust and Tracing Game, as the traders who benefitted from the use of *network* exploited their trusting clients. There was no evidence that trust affected the measurable transaction costs. Additional analyses showed that buyers detected cheats with other mechanisms than tracing. The traces showed more cheats than statistically possible when the envelopes were a random sample.

Section 5.3 presents the multi-agent model developed for the Trust and Tracing Game. This model has been tested and validated. It has been possible to validate the multi-agent simulation on an aggregate level against sessions with human participants. Hypotheses were formulated based upon observations of sessions. Each hypothesis could be confirmed in model runs.

Chapter 6 presents research with the second gaming simulation, called the Mango Chain Game (MCG). It was developed to study the bargaining power and revenue distribution among traders in the Costa Rican mango export chain. The MCG assessed what factors, including social structure, determined the bargaining power, what mode of organisation was used and how this influenced the revenue distribution between traders. The data collection combined data from a questionnaire among the participants with the actual behaviour in the game session. Five sessions were conducted with smallholders in the Costa Rica lowlands, resulting in 82 contracts. The results show that the bargaining power in the sessions was isomorphic to the real-world bargaining power of smallholders, multinationals and independent exporters. As expected, lower bargaining power on the part of the buyer (seller) resulted in higher revenue for the seller (buyer). In general, stakeholders with more bargaining power were able to take advantage of the other agents. Higher risk-aversion of the buyers and/or the sellers led to higher revenues for the other agents involved in the exchange relationship. In the same vein, long-term contracts in the buyer-seller relationship led to lower revenues (but also reduced risk) for sellers. The latter result was surprising, since contract choice appeared only to be significant for the seller's and not for the buyer's revenue equation. Mango producers turned out to be well aware of the fact that the type of markets in which they operate is mainly based on short-terms contracts. Not working with long-term contracts gave them the opportunity to remain flexible towards changes in demand and supply that they cannot control. Producers were trying to establish long-term relationships, but they could equally rely on repeat short-term contracts with the same partner. The latter type of contract tends to rely on trust or friendship, thus the network mode of organisation is at play here. Finally, real-world wealth appeared to have a significant impact on bargaining power.

Chapter 7 discusses the experiences with gaming simulation as a research method and draws conclusions from the combined results of the TTG and MCG. Section 7.1 discusses the experiences with gaming simulation to generate and test hypotheses. Methodologically there were some differences. Research with the TTG took more time than that with the MCG. The reasons for this difference can be found in the functions used and the number of variables. The TTG started with a broad scope, where the important variables coming from all four levels of the Williamson framework had yet to be found in the design cycle. In contrast to this, the MCG used an analytical model with fewer and theory-based variables. Attention is paid to validity and reliability. In summary, the process validity of both gaming simulations was the most important aspect, and psychological reality was required to get the process going. Both the TTG and MCG met these criteria in different ways. The MCG scored more positively on both the structural and predictive validity because of the closer resemblance of the supply network modelled with the real world and the use of real smallholders versus students. Care should be taken regarding claims about what is modelled.

The multi-agent simulation has been developed to perform sensitivity analyses of variable settings (loads). This project has not reached the point where variable settings selected with the MAS have been tested in a human session. Future research should make clear whether MAS really helps to increase efficiency by reducing the number of sessions needed through selection of interesting loads. This project has proven, however, that it is possible to develop and validate a multi-agent model of a gaming simulation.

The different types of data that the TTG and MCG can generate are hard to obtain using other methods. Gaming simulation is special in that the participants are exposed to a laboratory-like situation that isolates them from the real-world (trading) environment. In this laboratory environment, the attention of the participants can be focused on a particular problem, while retaining the full richness of human behaviour. Based upon the experiences presented in this book, gaming simulation can be positioned as a research method that facilitates a whole range of data collections. It is possible to acquire data before, during and after a session, enabling the coupling with questionnaires and interviews and actual observation of actions. It can analyse differences between participants in one session, testing for differences in backgrounds, or between session, testing for the effects of varying the load and situation in a session, or even the rules, roles, objectives and constraints of the gaming simulation itself. The combination of qualitative and quantitative analyses that is possible using gaming simulations makes the method a good candidate for research that requires both. In the methodology used in the MCG and TTG the first cycle (design cycle) was based upon a qualitative approach, while the hypotheses were tested in a quantitative empirical cycle.

Section 7.2 compares the theoretical conclusions of the Trust and Tracing Game with those of the Mango Chain Game and relates them to the main research question. The conclusions of the MCG are in line with the TTG conclusions insofar as the variables of social structure used in this book (trust, embeddedness, norms and values) clearly shape the organisation of the transactions. Both the TTG and MCG show that trust and embeddedness lead to the use of the network mode of organisation, but also to less revenue in the setting of these two gaming simulations. It seems that the network mechanism is essential in supply networks with independent traders but not in a way that directly leads to more revenues. The conclusions from the two gaming simulations are at odds with the leading paradigm in the literature on supply networks which says that trust and relationships are important for successful business. The conclusions are more in line with neo-classical economic theory, where companies use transactions as the rational result of considering price and product.

Section 7.3 presents ideas for future research divided into ideas for the domain and ideas for methodological improvements. Next, Section 7.4 presents the implications of the research for the domain of food supply chains and networks. The application of gaming simulation as (one of the) research method(s) can be of value for gathering data about the real behaviour of real participants in a simplified 'surrogate' environment, determined by the gaming simulation. In this book a laboratory for chain and network studies has been built. Section 7.5 looks at the implications for other domains and argues that the issues about trust, embeddedness and other variables from social structure can be found in most other business domains, thus providing opportunities for research with this method.

The chapter ends with some concluding remarks in Section 7.6, stating that this book showed as a proof-of-principle that gaming simulation is an excellent additional research method for controlled analysis of complex social systems. It also showed that the possibility to have a repeatable experiment within a controlled contextual setting provides insight into socio-economic behaviour in a way that can be approached from multiple bodies of theory. Staying within the framework of one body of theory cannot explain the full richness of human behaviour, thus links have to be made. In the current book the theoretical framework from new

Summary

institutional economics has been used and some first attempts have been made to link up with theory on culture, psychology and other theories in the social sciences. Future research could use gaming simulation as the research method of choice for true interdisciplinary research.

About the author

Name: Sebastiaan Arno Meijer
Address: Lombokstraat 22, 3531 RD Utrecht
Tel: +31(0)6-41618443
Mail: sebas@sebastiaanmeijer.com
Born: 6 January 1979 in Apeldoorn, The Netherlands
Matrimony: Single

Education:

2003 - 2008   PhD project on the organisation of transactions in supply networks. Development and test of gaming simulation methodology for assessing the social and relational aspects of modes of organisation.
Fieldwork included work in Costa Rica with local mango producers, sessions with student groups from all over the world and with groups of professionals. Oral paper presentations given at 11 international conferences. Course and workshop obligations completed as required by Mansholt Graduate School for the Social Sciences.

1998 - 2003   Free doctoral phase, Wageningen University built around human information usage on individual, company and chain level. Master thesis on design and implementation of a computer-based supply chain game and on generic modelling software for bio-mass production.

1997 - 1998   Propaedeuse Agricultural Engineering, Wageningen University

Extracurricular activities:

2004   Co-organisation of 8th International Workshop of IFIP WG 5.7
1994 - 1997   Driving force behind the students council at 'De Heemgaard'

Business experience:

1999- present Founder and owner of the company 'Basstec Scientific Noise'
2004 - 2005   Founder and co-owner (80%) Kabaal en Licht VOF.

Awards:

2004   Nominated for best demonstration of the 16th Belgian-Dutch conference on artificial intelligence, Groningen, the Netherlands

2006   Best paper award Production Planning and Control for Meijer, Hofstede, Beers and Omta: Trust and Tracing Game, learning about transactions and embeddedness in a trade network.

2008   Nominated for best paper award 8th International conference on chain and network management in agribusiness and food industry.

International chains and networks series

Agri-food chains and networks are swiftly moving toward globally interconnected systems with a large variety of complex relationships. This is changing the way food is brought to the market. Currently, even fresh produce can be shipped from halfway around the world at competitive prices. Unfortunately, accompanying diseases and pollution can spread equally rapidly. This requires constant monitoring and immediate responsiveness. In recent years tracking and tracing has therefore become vital in international agri-food chains and networks. This means that integrated production, logistics, information- and innovation systems are needed. To achieve these integrated global supply chains, strategic and cultural alignment, trust and compliance to national and international regulations have become key issues. In this series the following questions are answered: How should chains and networks be designed to effectively respond to the fast globalization of the business environment? And more specificly, How should firms in fast changing transition economies (such as Eastern European and developing countries) be integrated into international food chains and networks?

About the editor

Onno Omta is chaired professor in Business Administration at Wageningen University and Research Centre, the Netherlands. He received an MSc in Biochemistry and a PhD in innovation management, both from the University of Groningen. He is the Editor-in-Chief of The Journal on Chain and Network Science, and he has published numerous articles in leading scientific journals in the field of chains and networks and innovation. He has worked as a consultant and researcher for a large variety of (multinational) technology-based prospector companies within the agri-food industry (e.g. Unilever, VION, Bonduelle, Campina, Friesland Foods, FloraHolland) and in other industries (e.g. SKF, Airbus, Erickson, Exxon, Hilti and Philips).

Guest editors

Gert Jan Hofstede (1956) is an associate professor at Wageningen University, Social Sciences, Logistics, Decision & Information sciences group and at Delft University of Technology, Man-Machine interaction group. His interests include gaming simulation, trust and transparency in chains, cross-cultural communication, and the roots of these phenomena that can be found in our social biology. He publishes and lectures widely in these fields.

George Beers is working as a researcher in the field of information science especially in the domain of food chain management. Special fields of interest are strategic information management and research methodology. He works at Wageningen University.

Printed in the United States
by Baker & Taylor Publisher Services